Jan Taubert

ONDEX

Jan Taubert

ONDEX

A data integration framework for the life sciences

Südwestdeutscher Verlag für Hochschulschriften

Impressum/Imprint (nur für Deutschland/only for Germany)
Bibliografische Information der Deutschen Nationalbibliothek: Die Deutsche Nationalbibliothek verzeichnet diese Publikation in der Deutschen Nationalbibliografie; detaillierte bibliografische Daten sind im Internet über http://dnb.d-nb.de abrufbar.
Alle in diesem Buch genannten Marken und Produktnamen unterliegen warenzeichen-, marken- oder patentrechtlichem Schutz bzw. sind Warenzeichen oder eingetragene Warenzeichen der jeweiligen Inhaber. Die Wiedergabe von Marken, Produktnamen, Gebrauchsnamen, Handelsnamen, Warenbezeichnungen u.s.w. in diesem Werk berechtigt auch ohne besondere Kennzeichnung nicht zu der Annahme, dass solche Namen im Sinne der Warenzeichen- und Markenschutzgesetzgebung als frei zu betrachten wären und daher von jedermann benutzt werden dürften.

Coverbild: www.ingimage.com

Verlag: Südwestdeutscher Verlag für Hochschulschriften GmbH & Co. KG
Dudweiler Landstr. 99, 66123 Saarbrücken, Deutschland
Telefon +49 681 37 20 271-1, Telefax +49 681 37 20 271-0
Email: info@svh-verlag.de

Approved by: Bielefeld, Universität, Diss., 2010

Herstellung in Deutschland:
Schaltungsdienst Lange o.H.G., Berlin
Books on Demand GmbH, Norderstedt
Reha GmbH, Saarbrücken
Amazon Distribution GmbH, Leipzig
ISBN: 978-3-8381-2929-7

Imprint (only for USA, GB)
Bibliographic information published by the Deutsche Nationalbibliothek: The Deutsche Nationalbibliothek lists this publication in the Deutsche Nationalbibliografie; detailed bibliographic data are available in the Internet at http://dnb.d-nb.de.
Any brand names and product names mentioned in this book are subject to trademark, brand or patent protection and are trademarks or registered trademarks of their respective holders. The use of brand names, product names, common names, trade names, product descriptions etc. even without a particular marking in this works is in no way to be construed to mean that such names may be regarded as unrestricted in respect of trademark and brand protection legislation and could thus be used by anyone.

Cover image: www.ingimage.com

Publisher: Südwestdeutscher Verlag für Hochschulschriften GmbH & Co. KG
Dudweiler Landstr. 99, 66123 Saarbrücken, Germany
Phone +49 681 37 20 271-1, Fax +49 681 37 20 271-0
Email: info@svh-verlag.de

Printed in the U.S.A.
Printed in the U.K. by (see last page)
ISBN: 978-3-8381-2929-7

Copyright © 2011 by the author and Südwestdeutscher Verlag für Hochschulschriften GmbH & Co. KG and licensors
All rights reserved. Saarbrücken 2011

TABLE OF CONTENTS

1 Introduction 1
2 Background and related work 7
   2.1 Principles of data integration 7
      2.1.1 Link integration and hypertext navigation 7
      2.1.2 Data warehouses 8
      2.1.3 View integration and mediator systems 9
      2.1.4 Workflows 10
      2.1.5 Mashups 10
   2.2 Previous work 11
      2.2.1 SEMEDA 12
      2.2.2 PROTON 15
   2.3 Survey of current data integration systems 17
      2.3.1 Visual Knowledge and BioCAD 19
      2.3.2 Biozon 21
      2.3.3 BNDB / BN++ 22
      2.3.4 STRING 24
      2.3.5 NeAT 25
   2.4 Conclusion 26
3 Requirements 29
   3.1 Current situation 29
   3.2 Challenges for data integration 30
   3.3 Comparison with previous and related work 33
   3.4 Comparison of approaches to data integration 35
   3.5 Requirements for ONDEX 37
4 Methods and principles 41
   4.1 ONDEX integration data structure 41
      4.1.1 Motivation 41
      4.1.2 Semantics in ONDEX 43
      4.1.3 Semantics of nodes 44
      4.1.4 Semantics of edges 46
      4.1.5 Provenance 49
      4.1.6 References and synonyms 50
      4.1.7 Generalised data structure 52
      4.1.8 Context 53
      4.1.9 Definition ONDEX integration data structure 54

Contents

- 4.1.10 Discussion 55
- 4.2 Data alignment 57
  - 4.2.1 Motivation 57
  - 4.2.2 Methods 59
  - 4.2.3 Results 70
  - 4.2.4 Discussion 79
- 4.3 Exchanging integrated data 81
  - 4.3.1 Motivation 81
  - 4.3.2 Requirements for exchanging integrated data sets 82
  - 4.3.3 The OXL format 85
  - 4.3.4 Discussion 98
  - 4.3.5 Outlook 100

5 Design and implementation 101
- 5.1 System design 101
  - 5.1.1 Knowledge modelling and domain independence 103
  - 5.1.2 Formulating a consensus domain model in biology 105
  - 5.1.3 Populating the domain model 108
  - 5.1.4 Data filtering and knowledge extraction 113
  - 5.1.5 Workflows 115
- 5.2 Implementing integration data structure 117
  - 5.2.1 Encapsulation 117
  - 5.2.2 Inheritance 117
  - 5.2.3 Association 118
  - 5.2.4 Polymorphism 119
  - 5.2.5 Aspect oriented development 121
  - 5.2.6 Graph implementations and persistency 121
- 5.3 Graph querying and information retrieval 123
  - 5.3.1 Views based on semantics 123
  - 5.3.2 Indexing and querying 126

6 Use cases 129
- 6.1 Improving genome annotations for *Arabidopsis thaliana* 129
  - 6.1.1 Motivation 129
  - 6.1.2 Data integration approach 131
  - 6.1.3 Data integration exemplar 132
  - 6.1.4 Data integration pipeline 133
  - 6.1.5 Evaluation of annotation methods 137
  - 6.1.6 Discussion 140

6.2 Prediction of potential pathogenicity genes in *Fusarium graminearum* 143
    6.2.1 Motivation 143
    6.2.2 Methods 144
    6.2.3 Results 146
    6.2.4 Discussion 149
6.3 Constructing a consensus metabolic network for *Arabidopsis thaliana* 151
    6.3.1 Motivation 151
    6.3.2 Methods 152
    6.3.3 Results 156
    6.3.4 Discussion 160
6.4 Analysis of social networks 161
    6.4.1 Motivation 161
    6.4.2 Methods 162
    6.4.3 Results 165
    6.4.4 Discussion 169
6.5 ONDEX SABR project and its applications 171
7 Conclusion and outlook 177
    7.1 Summary 177
    7.2 Design decisions 183
    7.3 Addressing the challenges 185
    7.4 Comparison with related work 191
    7.5 Outlook 193
8 Glossary 197
    8.1 Definitions 197
        8.1.1 Graph theory 197
        8.1.2 Properties on relations 198
        8.1.3 Object-oriented development and UML 198
        8.1.4 Aspect-oriented software development 199
    8.2 Knowledge representation 201
        8.2.1 Semantic network 201
        8.2.2 Conceptual graphs 202
        8.2.3 Ontology 203
        8.2.4 Contexts 203
        8.2.5 Domain knowledge 204
        8.2.6 Domain modelling 204
        8.2.7 Hierarchy and taxonomies 205
        8.2.8 Controlled vocabulary 205

Contents

8.3 Terms used in ONDEX	207
9 Appendix	211
9.1 List of data formats supported by ONDEX	211
9.1.1 Import	211
9.1.2 Export	212
9.2 Data integration methods in ONDEX	213
Table of figures	217
References	223

Acknowledgements

I would like to thank all my supervisors Prof. Dr. Ralf Hofestädt, Dr. Jacob Köhler and Prof. Dr. Christopher Rawlings for their valuable input and support. Without it, this work would have never been completed. In particular, I would like to thank Prof. Dr. Ralf Hofestädt for having me as an external member of the AG Bioinformatics at University of Bielefeld since my undergraduate studies, Dr. Jacob Köhler as my early mentor and his invaluable help, ideas and advice with creating the ONDEX system and Prof. Dr. Christopher Rawlings for his suggestions and comments on my work as well as for allowing me enough free time to finish this thesis aside my employment within the Centre for Mathematical and Computational Biology at Rothamsted Research. I also would like to thank Prof. Dr. Ipke Wachsmuth and Dr. John McCrae for agreeing to be members of my examination board in addition to Prof. Dr. Ralf Hofestädt and Prof. Dr. Christopher Rawlings.

Furthermore I would like to thank my fellow students Keywan Hassani-Pak, Matthew Hindle, Berend Hoekman, Artem Lysenko, Robert Pesch, Klaus Peter Sieren, Jochen Weile and Rainer Winnenburg for their help in implementing some of the presented ideas in ONDEX and with preparing the manuscripts of the publications which contributed to this thesis.

Last but not least, I would like to thank all my family, especially my wife Natalia and daughter Helena, for all their patience, understanding and support during this long process.

1 Introduction

Over the last decade biological research has changed completely. The reductionism approach of studying only a few biological entities at a time in the past is being replaced by the study of the biological system as a whole today. Systems Biology [1] seeks to understand how complex biological systems work by looking at all parts of biological systems, how they interact with each other and form the complete whole. Systems Biology can be seen as a cycle (see Figure 1) consisting of the following steps:

- having a testable hypothesis about a biological system
- conducting experimental validation of hypothesis
- capturing and analysis of experimental results (usually 'omics data)
- gain new insights (data) about a biological system from analysis results
- refine model about a biological system to derive new hypothesis

This process requires that existing biological knowledge (data) is made available to support on the one hand the analysis of experimental results (see Figure 1a) and on the other hand the construction and enrichment of models for Systems Biology (see Figure 1b).

Effective integration of biological knowledge from databases scattered around the internet and other information resources (for example experimental data) is recognized as a pre-requisite for many aspects of Systems Biology research and has been shown to be advantageous in a wide range of use cases such as the analysis and interpretation of 'omics data [2], biomarker discovery [3] and the analysis of metabolic pathways for drug discovery [4]. However, systems for the integration of biological knowledge have to overcome several challenges. For example, biological data sources may contain similar or overlapping coverage and the user of such systems is faced with the challenge of generating a consensus data set or selecting the "best" data source. Furthermore, there are many

1 Introduction

technical challenges to data integration (see 3.2 "Challenges for data integration"), like different access methods to databases, different data formats, different naming conventions and erroneous or missing data.

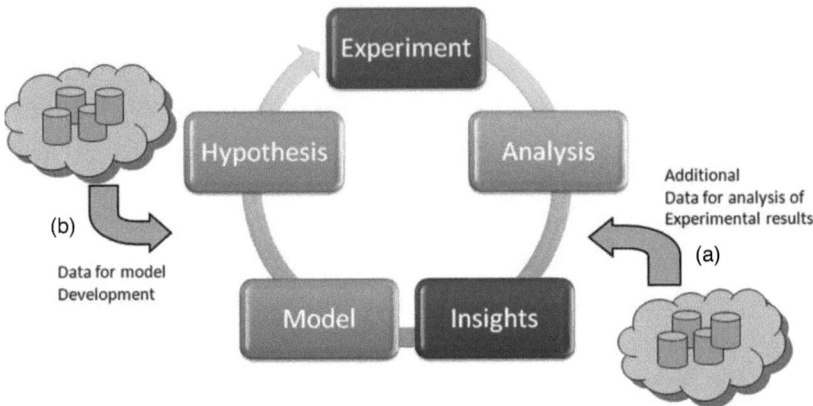

FIGURE 1 SYSTEMS BIOLOGY CYCLE OF EXPERIMENT, ANALYSIS, INSIGHTS, MODEL AND HYPOTHESIS TOGETHER WITH REQUIREMENTS FOR LARGE DATA FOR ANALYSIS OF EXPERIMENTAL RESULTS AND MODEL DEVELOPMENT

To address these challenges and enable effective integration of biological knowledge in support of Systems Biology research, the ONDEX system [2, 5-7] which is presented in this thesis was created. The ONDEX system provides an integrated view across biological data sources with the aim to enable the user to gain a better understanding of biology from integrated knowledge. ONDEX is supported by BBSRC (http://www.bbsrc.ac.uk/) as part of the System Approaches to Biological Research initiative (SABR) and is now mainly being developed at Rothamsted Research, Manchester University and Newcastle University. The first ONDEX prototype was developed at University of Bielefeld.

ONDEX (see Figure 2) uses a three steps approach to address the outlined challenges, namely import of data from data sources ("Data Input", left), identifying overlapping or similar data across different data sources ("Data Integration", middle) and analysis of the resulting integrated datasets to reveal new biological insights ("Data

Analysis", right). This thesis presents the most interesting aspects of the ONDEX system which are highlighted in Figure 2 in more detail; all other parts of the ONDEX system will only be briefly described when appropriate. Although the selection of work presented in this thesis appears to be distinct, many contributions to other parts of the ONDEX system have also been made over time.

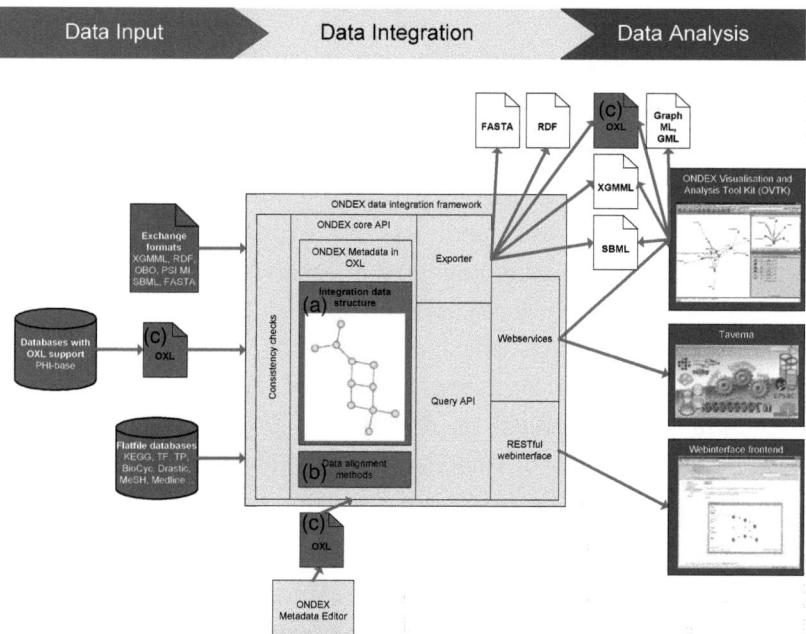

FIGURE 2 OVERVIEW OF ONDEX SYSTEM USING A THREE STEP APPROACH OF 1) DATA INPUT (LEFT), 2) DATA INTEGRATION (MIDDLE) AND 3) DATA ANALYSIS (RIGHT). HIGHLIGHTED PARTS WILL BE THE MAIN TOPICS OF THIS THESIS; ALL OTHER PARTS WILL ONLY BE BRIEFLY DESCRIBED.

Chapters 2 and 3 of this thesis provide an overview of the field of research and layout the requirements for the ONDEX system. Chapter 4 follows the structure highlighted in Figure 2. Chapter 4 breaks down into the integration data structure (Figure 2a) which is used to accommodate data from different sources as described in 4.1 "ONDEX integration data structure", followed by introducing and evaluating approaches to identify equivalent or similar entities across different data sources in 4.2 "Data alignment" (Figure 2b) and the transport or exchange of datasets as a prerequisite to data analysis (Figure 2c) discussed in 4.3 "Exchanging integrated data". More general aspects of the use and implementation of the ONDEX system are presented in Chapter 5 "Design and implementation", which also addresses the problems and challenges during data import in 5.1.3 "Populating the domain model" (left side in Figure 2). The last part of this thesis presents a selection of different data analysis scenarios (right side in Figure 2) in Chapter 6 "Use cases".

These data analysis scenarios or use cases have been selected to illustrate some of the applications of ONDEX in Systems Biology research. Three major use cases are presented in this thesis:

> In Figure 1a it is illustrated that integrated biological data is required for the analysis of experimental results. In transcriptomics experiments, it is essential to know the correct gene and its functional annotation for all the gene probes on a microarray gene chip. Section 6.1 "Improving genome annotations for *Arabidopsis thaliana*" uses the model plant *Arabidopsis thaliana* as an example organism to show how genome annotations can be successfully enriched utilizing an automated pipeline in ONDEX. An evaluation of the annotation results is given.

> With the dawn of high-throughput sequencing technology the number of newly sequenced genomes is rapidly increasing. Using insights gained by comparative genomics can support the understanding of biology of novel sequenced organisms as shown in Section 6.2 "Prediction of potential pathogenicity genes in *Fusarium graminearum*".

> Building models in Systems Biology (see Figure 1b) can benefit from an integrated view across multiple data sources. These data sources might not always be complementary and could even contain contradicting information. Therefore it is necessary to construct a consensus network across data sources as shown in Section 6.3 "Constructing a consensus metabolic network for *Arabidopsis thaliana*".

As part of the ONDEX SABR project (http://www.ondex.org) several Systems Biology applications using ONDEX have been developed, which are summarized in 6.5 "ONDEX SABR project and its applications".

This thesis finishes with Chapter 7 "Conclusion" about the achievements and contributions made in comparison to existing work. These include definition of a graph-based integration data structure, construction of a consensus domain model for biological pathway data, methods for addressing the challenges of syntactic and semantic heterogeneities in data sources, presentation and evaluation of novel algorithms for identifying equivalent or related entities across data sources and an exchange format for integrated datasets. Further developments of ONDEX towards an Intellectual Access Ramp [8] for biologists based on Web 2.0 technology are mentioned in Section 7.5 "Outlook".

1 Introduction

2 Background and related work

2.1 Principles of data integration

Recent reviews of common data integration principles for bioinformatics can be found in [9-13]. Here the most commonly used approaches to data integration are highlighted:
- Link integration and hypertext navigation
- Data warehouses
- View integration and mediator systems
- Workflows
- Mashups

A comparison of these approaches is given in Chapter 3 "Requirements" concluding in some design requirements for the ONDEX data integration system.

2.1.1 Link integration and hypertext navigation

Most public databases provide cross-references to entries in other databases [14]. Usually these cross-references are accession numbers (stable database specific identifiers, which are generated when a new entry is added to a database). They are made available on the webpage of the database so that the user can surf across datasets by following hyperlinks.

The use of ontology and identity authorities is necessary to enable the cross-referencing between databases. In [9], SRS [15], Entrez [16] and Integr8 [17] are given as examples of link integration systems. These systems provide webpage portals and are based on keyword indexing to establish and maintain the cross-database reference network.

Depending on the complexity of the indexing system (for example awareness of different data types, use of ontologies for query construction etc.), a user can ask more or less sophisticated queries. As access to major parts of the data is still located with the originating data source, the results of a user query can be kept up-to-date. On the other hand link integration systems have to rely on the availability of the originating data source.

Cooperation of service providers is required to make link integration work well. It has to cope with problems of name clashes, term ambiguities and updates of the schemas of the linked data sources. The integration process does not have an underlying domain model, thus it is model independent. It is more inter-linking of data sources than data integration.

2.1.2 DATA WAREHOUSES

In contrast to link integration where the majority of data remains with the data provider, data sets are extracted from the original data source, cleaned and transformed into a pre-determined domain or data model. Ideally this data model is more or less stable once it has been filled with data. The integrated data is stored and queried as a single integrated resource. This is true data integration from many resources into one, long-standing resource.

Goble et al. [9] emphasise the popularity of data warehouses for certain communities (for a particular species), for example e-Fungi [18] for fungal genomes or GIMS [19] for *Saccharomyces cerevisiae* (yeast) genomes, or disciplines (genomics, proteomics etc.), for example ATLAS [20] for genomics data in the biomedical domain or Columba [21] for proteomics data. Furthermore data warehouses are commonly found within enterprises for in-house data gathering. The toolkits which facilitate the construction of such in-house data warehouses listed include IBM's Websphere Information Integrator [22], GMOD [23], BioMART [24] or BioWarehouse [25].

Although widely used, this approach comes with high initial activation costs as the data model has to be created and populated with the data sources. The design of the data model is often an act of faith that the data warehouse will be needed in the way it has been built. Furthermore, keeping track of provenance for data contained in the data warehouse is not trivial.

Data warehouses find it hard to cope with changes in requirements or data as the data model used is often fixed or hard to change. They are commonly decoupled from their data providers. This results in high maintenance costs to keep the content of data warehouses up to date with changes in their sources. Although data warehouses are popular, they find it hard to cope with a world in flux.

2.1.3 VIEW INTEGRATION AND MEDIATOR SYSTEMS

In contrast to data warehouses, where data is kept locally, view integration and mediator systems leave their data in its source databases. An environment is built which presents all the connected databases as one single database. Such systems typically consist of three elements: wrappers, an integration layer and a query interface. The wrappers provide uniform access to the data sources and ensure that the content of the view model is always up to date. The integration layer decomposes user queries to be processed by the wrappers and finally integrates the query results before the result is returned to the user via the query interface. The result is a kind of "virtual warehouse" maintained by a mediator processor and a series of mappings from the integrating model to source databases.

This approach has enjoyed wide acceptance in the life sciences. Goble et al. [9] list for example Biozon [26], TAMBIS [27], Kleisli [28], Medical Integrator [29] and ComparaGrid [30]. Some of these systems like TAMBIS or ComparaGrid use ontology as global schema with reasoning to overcome limitations of the underlying data sources. Unfortunately view integration and mediator systems carry the same high initial cost of designing the global integration model as a data warehouse. The models are often complex and hard to adapt. The wrappers and mappings involve high maintenance costs and might become fat and brittle in the face of unreliable and dynamic resources; and the whole environment is complex, both for the developer and the user. They also tend to be as slow as the slowest source.

2.1.4 WORKFLOWS

As semantic web technologies become more and more mature, service and database providers adapt these technologies and enable access to their resources via web services. Workflows use these web services to build ad-hoc data integration solutions which are lightweight, transparent and can be shared with the community, see MyExperiment [31].

Workflow systems vary. Some, like InforSense [32] and Pipeline Pilot [33], expect strong data type compliance so that the workflow effectively builds an on-the-fly data warehouse, or they presume a common data model. Others, like Taverna [34], have an open type system where the data passed between the workflow steps is "massaged" into shape by special "shim" processors and it is part of the responsibility of the workflow to build a data model if required [9]. Furthermore, workflows can be used to populate data warehouses.

Nevertheless, workflows are flexible and adaptable. They do not require the pre-existence of a single data model. Workflows are able to cope with unreliable data resources by catering for service substitutions and faults. Integration details are typically exposed as part of the workflow, so the evidence for data integration is made directly visible.

2.1.5 MASHUPS

Mashups provide functional capability or content using ideas based on Web 2.0. They are getting increasingly used in the life sciences. Mashups combine data from more than one web based data source to create a new one. A simple example would be overlaying the global spread of avian flu as reported by Nature News with geospatial data of Google Earth. Mashups emphasize the role of the user in creating a specific, light touch, on-demand integration, following the mantra "just in time, just enough" design. Thus Web 2.0 Mashups are built upon the existence of common *de facto* APIs alongside a collection of light-weight tools techniques for rapid and agile deployment, so that small specific solutions can be built for particular problems.

Goble et al. [9] lists the Distributed Annotation Service (DAS) [35] as an early form of a Mashups service. DAS based services can be combined using Mashups. An example given by the authors is the UniProt DASTY client, which combines 26 DAS servers to overlay third party information about sequence annotations on top of the display of a UniProt sequence [36]. The prime activity of Mashups is aggregation of data, not fully integration of data. Similar to link integration, Mashups depend on having some kind of cross-database references or common touch-point between the various data. They are just as vulnerable as all other data integration approaches to identity clashes and concept ambiguities.

2.2 PREVIOUS WORK

The first prototype of the ONDEX system has been developed by Jacob Köhler and colleagues at University of Bielefeld in 2004 before development was transferred to Rothamsted Research in 2005. Many diploma students from University of Bielefeld have since worked on ONDEX until the ONDEX SABR project started in April 2008. At the start of the ONDEX SABR project the current ONDEX software was created anew taking ideas of the previous versions into account.

Ideas taken from the federated data integration system SEMEDA [37] developed by Jacob Köhler at University of Bielefeld, have been incorporated into the ONDEX system from the outset. The main methods used in SEMEDA have been briefly described in the next section to show how these ideas are further developed in the work presented in this thesis. SEMEDA and the ONDEX system share the same main idea of "ontology based semantic integration of biological databases" to overcome the semantic heterogeneities in data sources.

Proton developed by Andreas Freier at University of Bielefeld is an example of how data integration can contribute to the construction of metabolic networks. These networks can then be used for simulation and understanding of biological processes. This approach has beside others inspired work in ONDEX to create a consensus metabolic network for *A. thaliana* as a use case (see 6.3 "Constructing a consensus metabolic network for *Arabidopsis thaliana*").

Both systems, Proton and SEMEDA, are no longer in active use. However, some of their functionality and design has been influencing the current ONDEX software.

2.2.1 SEMEDA

SEMEDA: ontology based semantic integration of biological databases [37, 38] was developed at the University of Bielefeld by Jacob Köhler et al. Some of the ideas developed in SEMEDA have had an influence on the ONDEX data integration framework presented in this thesis. Here the main concepts behind SEMEDA are introduced.

SEMEDA tried to address issues related to semantics after the technical problems of data integration had been overcome by using the BioDataServer [39] as a consolidated view on source databases. Köhler et al. [37] list the use of different terms for the same things, different names for equivalent database attributes and missing links between relevant entries in different databases as examples for remaining issues related to semantics. SEMEDA uses a federated approach (see 2.1.3 "View integration and mediator systems") to data integration and is provided via a web-interface and a database API.

In a first step SEMEDA assigns semantics to database attributes, see Figure 3 taken from [37], by mapping them to ontology concepts, thus overcoming differences in attribute naming across different databases. The ontology used should be formal with respect to the implementation of a transitive 'is-a' hierarchy, which connects all concepts. Since database attributes should be semantically defined to be as specific as possible, no database attributes is semantically defined as general top-level concepts such as 'thing'.

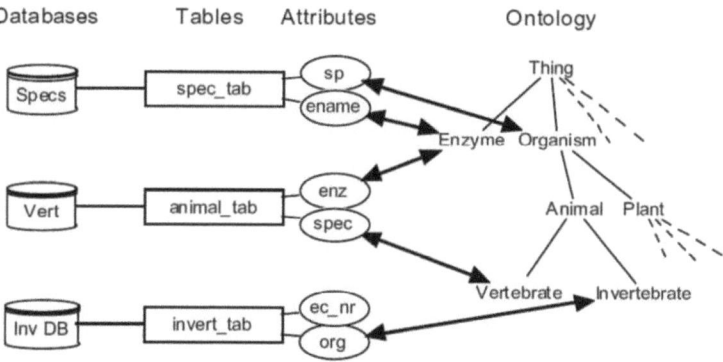

FIGURE 3 DATABASE ATTRIBUTES CAN BE DEFINED BY LINKING THEM TO ONTOLOGY CONCEPTS (THICK ARROWS), 'ENAME' AND 'ENZ' ARE DEFINED IN THIS EXAMPLE AS THE SAME CONCEPT 'ENZYME', I.E. THEY BOTH CONTAIN ENZYME NAMES. 'ORG' CONTAINS ONLY INVERTEBRATES, WHEREAS 'SPEC' CONTAINS VERTEBRATES.

In a second step semantics for the attribute values are defined by using controlled vocabularies as data types for attributes. By listing synonymous concepts between controlled vocabularies in a *translation list*, it is possible to relate entries between databases that use different terms for the same thing, see Figure 4 taken from [37]. This can overcome differences in naming at the database entries level.

2 Background and related work

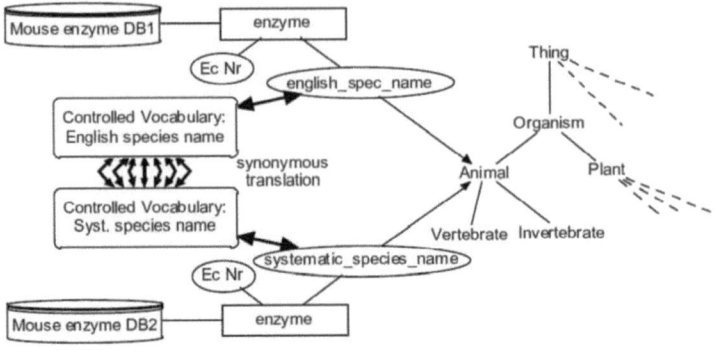

FIGURE 4 BY MAPPING SYNONYMOUS CONCEPTS OF CONTROLLED VOCABULARIES, IT IS POSSIBLE TO RELATE DATABASE ENTRIES THAT USE DIFFERENT TERMS FOR THE SAME THINGS.

Once related entries between databases have been identified, it is now possible to use these cross-references to automatically generate database links as well as to automatically derive which database tables can be joined, see Figure 5 taken from [37]. The query 'animal:mouse' as an example, which would in case of Figure 5 find the EC number of mouse enzymes. By using semantic cross-references, a system can automatically generate links to other database tables that contain further information about EC numbers. In the example, additional information can therefore be found in the 'enzymes' database.

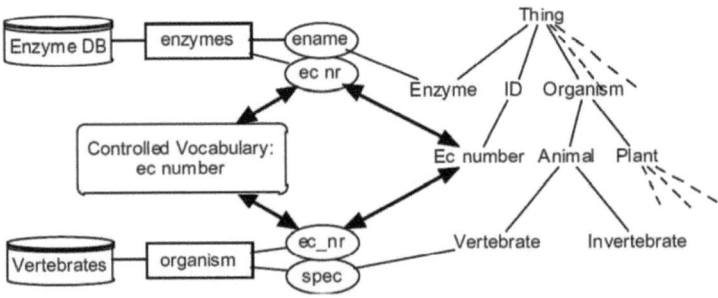

FIGURE 5 DATABASE ATTRIBUTES WHICH ARE DEFINED AS THE SAME CONCEPT AND SHARE THE SAME CONTROLLED VOCABULARY AS THEIR DOMAIN CAN BE USED FOR CROSS-REFERENCING BETWEEN DATABASE ATTRIBUTES.

In Figure 6 taken from [37] the different steps of performing a query using SEMEDA are depicted. This interface enables users to query databases without requiring knowledge of the structure or any technical details about the data sources. In addition, it guides users to relevant databases, and provides cross-references from query results to other relevant databases.

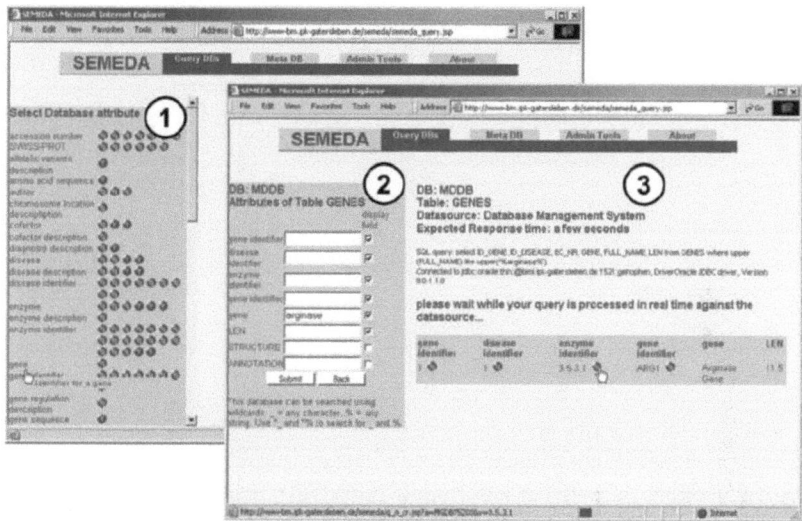

FIGURE 6 SEMEDA'S DATABASE QUERY INTERFACE. (1) ALL CONCEPTS FOR WHICH DATABASE ATTRIBUTES EXIST ARE LISTED. EACH OF THE ROUND ICONS REPRESENTS A DATABASE TABLE THAT CONTAINS AN ATTRIBUTE FOR THE CONCEPT. (2) AFTER SELECTING ONE OF THE ICONS, AN APPROPRIATE FORM OF THE RESPECTIVE TABLE MAY BE USED TO QUERY THE DATABASE. (3) THE RESULT SET IS EXTENDED WITH ICONS THAT ARE USED TO CROSS-REFERENCE OTHER DATABASES.

2.2.2 PROTON

PROTON (Prototyping Object Networks) [40] is an object-oriented system for modelling, integration and analysis of gene controlled metabolic networks by reconstructing biochemical systems from molecular databases developed at University of Bielefeld. This approach includes models for the design of biochemical networks as objects, processes and systems. At the start, information from dynamically adopted data sources is extracted and fused as molecular objects. In a second step, biochemical processes are identified interactively and large-scale state charts are automatically constructed from integrated data. Finally, networks of biochemical reactions are detected in the state charts and they are used to step wisely build the structure of dynamic systems.

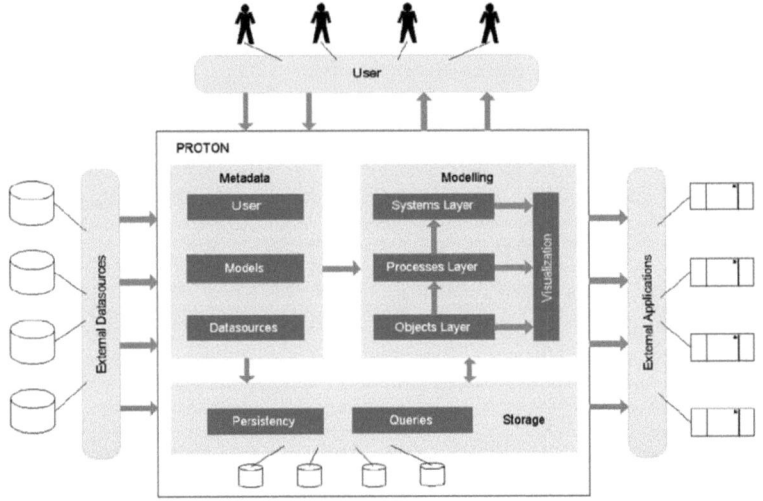

FIGURE 7 PROTON SYSTEM OVERVIEW DEPICTING THE USE OF METADATA (MIDDLE LEFT) FOR THE MODELLING AND VISUALISATION PROCESS (MIDDLE RIGHT) AND THE STORAGE INTERFACES OF THE SYSTEM (BOTTOM).

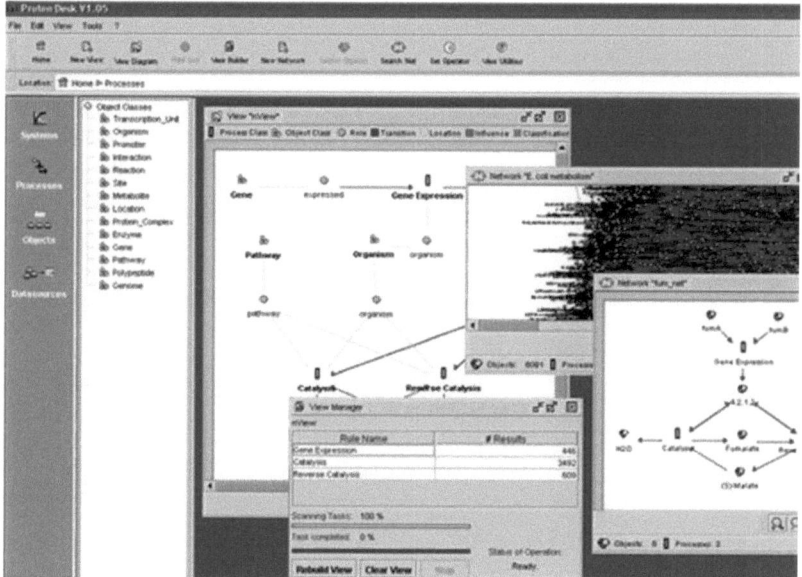

FIGURE 8 VISUALISATION OF INTEGRATED DATA IN PROTON SHOWING DIFFERENT VIEWS ABOUT METABOLIC NETWORKS.

Figure 7 shows a system overview for PROTON. Data from external sources is load via adapters and a mapping between data sources is created using a graphical editor driven by automated information extraction. The integrated datasets can be visualised as shown in Figure 8 and used for simulation of for example metabolic networks.

2.3 SURVEY OF CURRENT DATA INTEGRATION SYSTEMS

Several data integration systems for use in biology and related domains are in use today. Some of them use a generic approach to answer a wide range of biological questions. Others are more limited in their scope and application domain. These systems are based on principles such as Link integration and hypertext navigation, Data warehouses, View integration and mediator systems, Workflows and Mashups (see 2.1 "Principles of data integration") and are reviewed in more detail in [12] and [13].

2 Background and related work

Software tools that solve aspects of the data integration problem are being developed for some time. The early approaches, which produced popular software such as SRS [15] uses indexing methods to link documents or database entries from different databases and provide a range of text and sequence-based search and retrieval methods for users to assemble related data sets. The methods used by SRS (and related tools) address what has been described as the technical integration challenge.

More recently, data integration approaches are developed that "drill down" into the data and seek to link objects at a more detailed level of description. Many of these approaches exploit the intuitively attractive representation of data as graphs or networks with nodes representing things and edges representing how they are related. For example a metabolic pathway could be represented by a set of nodes identifying the metabolites linked by edges representing enzymatic reactions. Data integration systems that exploit graph based methods include PathSys [41] or BN++ [42] and the ONDEX system [43]. Both BN++ and ONDEX are available as open source software.

The methods presented in the course of this thesis are implemented as part of the ONDEX system. The related work presented here and in the previous chapter has been selected to highlight several different aspects which will be referred to during the principles and methods for the ONDEX system and will be used to compare with the motivation behind the work in this thesis.

The Visual Knowledge and BioCAD software tools provide good examples for how semantic networks (see 8.2.1 "Semantic network") can be used for representing biological knowledge. The definition of the integration data structure of ONDEX (see 4.1 "ONDEX integration data structure") has been inspired by this use of semantic networks in the biology domain.

Biozon is a data warehouse which includes additional derived information, such as sequence similarity or function prediction, between data entries. STRING shows that multiple information sources can be combined to provide evidence for the relationship between proteins. Similar to Biozon and STRING, ONDEX facilitates the integration of other derived information between data entities (see 4.2.2.3 "Other data integration methods"). Such information has been successfully used to improve genome annotation in a use case (see 6.1 "Improving genome annotations for *Arabidopsis thaliana*").

BNDB with BN++ is the most similar system to ONDEX in terms of system design and methodology. It is described here to be able to compare the two systems against each other in the last chapter of this thesis (see 7.4 "Comparison with related work").

The NeAT toolkit highlights how graph analysis applied to biological networks can help to reveal new insights. Furthermore it is a good example of providing such functionality via a webpage. Providing analysis functionality of integrated data using ONDEX via a webpage is considered as part of the Outlook section (see 7.5 "Outlook").

2.3.1 VISUAL KNOWLEDGE AND BIOCAD

In the combined systems of Visual Knowledge and BioCAD a semantic network (SN) approach has been used to develop and implement a model of cell signalling pathways [44]. A semantic network is a method to represent information or knowledge by nodes and edges in a graphical form, where a node represents a concept and an edge represents a relationship [45]. Hsing et al. implemented a system based on the combination of the Visual Knowledge [5] application development environment and its extension BioCAD [46]. The BioCAD software provides tools for managing large-scale biological data and for visualizing and editing biological pathways and networks. The authors conclude that the semantic network is an effective method to model cell signalling pathways. The semantic model allows proper representation and integration of information on biological structures and their interactions at different levels [44].

2 Background and related work

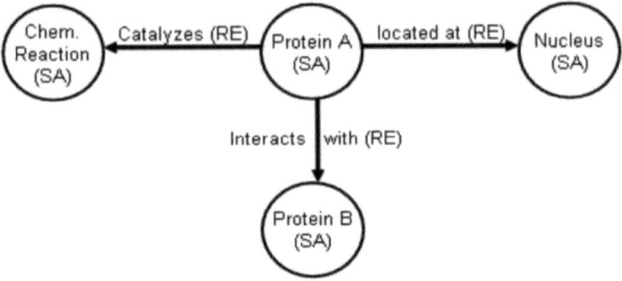

FIGURE 9 EXAMPLE OF A SEMANTIC NETWORK. CHARACTERISTICS AND BEHAVIOURS OF A SEMANTIC AGENT (SA) ARE DEFINED BY ITS RELATIONSHIPS (RE) WITH OTHER AGENTS.

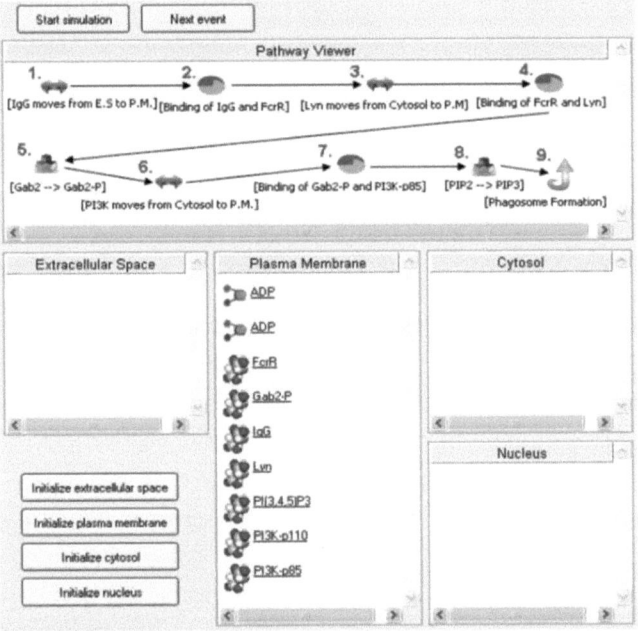

FIGURE 10 SCREENSHOT OF SEMANTIC NETWORK (SN) SIMULATOR. UPPER PART DEPICTS THE PATHWAY VIEW; LOWER PART REVEALS THE LOCATION OF REACTANTS.

Figure 9 shows an example of a semantic network as proposed by Hsing et al. [44]. The characteristics and behaviours of a semantic agent (SA) represented as a node in Figure 9 are defined by its relationships (RE) with other agents represented by arrows in Figure 9. Figure 10 is a screenshot of the SN simulator in the final state of the simulation. The upper part depicts the pathway view, whereas the lower part reveals the location of reactants.

2.3.2 BIOZON

Biozon is a unified biological resource for DNA sequences, proteins, complexes and cellular pathways [12]. It uses a single extensive and tightly connected graph schema wrapped with a hierarchical ontology of documents and relations. Beyond warehousing existing data, Biozon computes and stores novel derived data, such as similarity relationships and functional predictions. The integration of similarity data allows propagation of knowledge through inference and fuzzy searches. The database is accessible through a web interface that supports complex queries, "fuzzy" searches, data materialisation and more [26].

Figure 11 shows the front-page of biozon.org. The user can enter search terms or use more advanced search techniques like sequence searches. The results are presented and ranked in a tabular view.

2 Background and related work

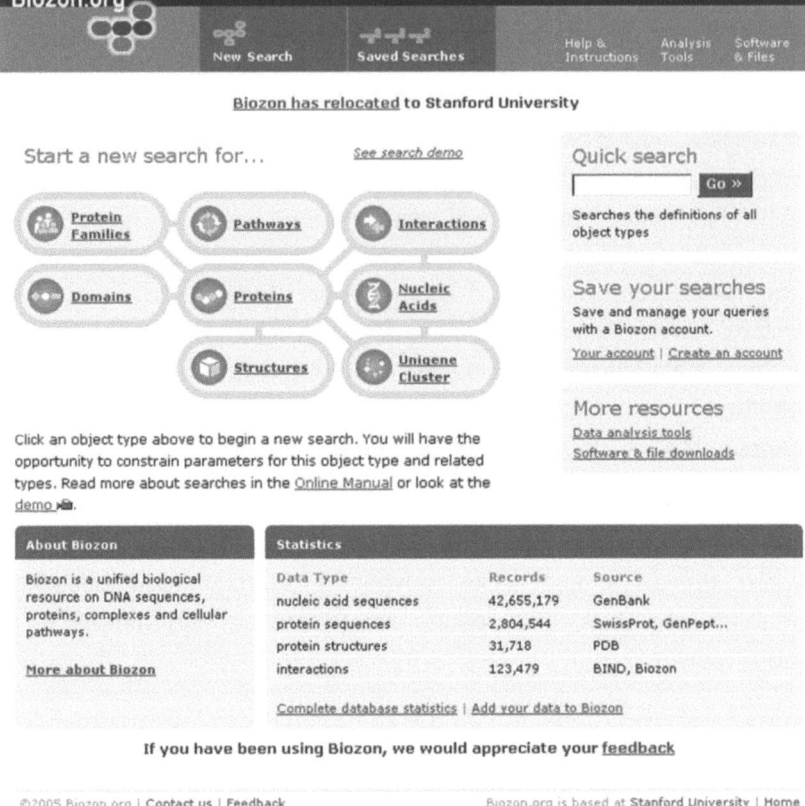

FIGURE 11 FRONT-PAGE OF BIOZON. CENTRE: SHOWING DIFFERENT OPTIONS TO START A SEARCH. BOTTOM RIGHT: SOME STATISTICS ABOUT THE INTEGRATED DATASETS AND DATA SOURCES.

2.3.3 BNDB / BN++

BNDB / BN++ [47] is a biochemical network library for analysing and visualising complex biochemical networks and processes. The software library is designed to simplify the implementation of tools for answering a broad range of interesting biochemical questions. It is available free of charge under the GPL for Linux, Windows and MacOS. BN++ uses an object-oriented data model (BioCore) provided as C++ and Java implementations. The data warehouse has a graphical user-interface (Biological Network Analyzer - BiNA).

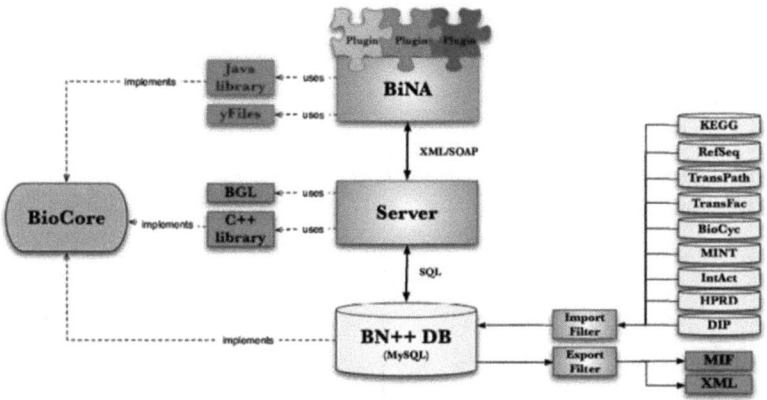

FIGURE 12 OVERVIEW ARCHITECTURE BNDB / BN++. THREE LAYERS FROM BOTTOM TO TOP: DATABASE LAYER WITH IMPORTERS (RIGHT) FOR DIFFERENT DATA SOURCES, A WEB SERVER PROVIDING A SOAP INTERFACE (MIDDLE) AND FRONT-END APPLICATION BINA (TOP) WITH PLUG-INS.

Figure 12 shows the modularized architecture of BNDB / BN++. Import filters for a range of biological databases exist to populate the object model of BNDB (MySQL). A server XML/SOAP interface connects BNDB to the visualisation front end BiNA depicted in Figure 13.

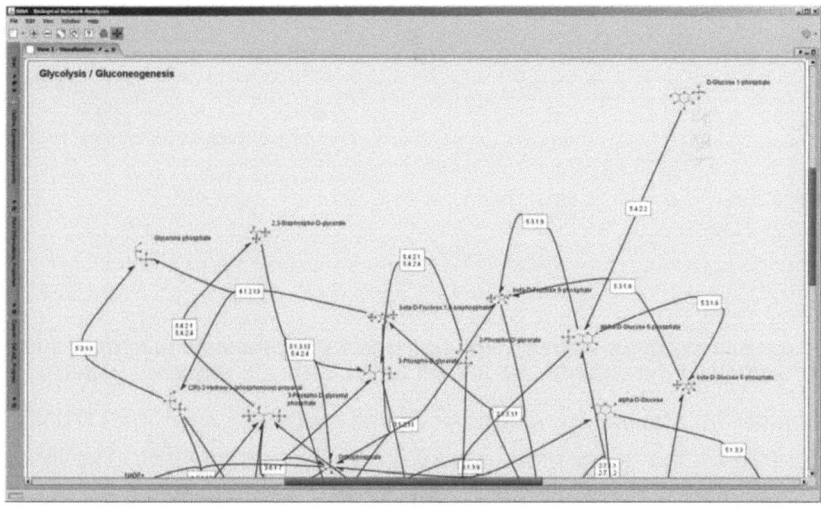

FIGURE 13 VISUALISATION FRONT-END BINA SHOWING AN EXAMPLE METABOLIC NETWORK WITH EC NUMBERS ATTACHED TO EDGES AND METABOLITES REPRESENTED ON NODES.

2.3.4 STRING

The database and web-tool STRING (Search Tool for the Retrieval of Interacting Genes/Proteins) is a meta-resource that aggregates most of the available information on protein–protein associations, scores and weights it, and augments it with predicted interactions, as well as with the results of automatic literature-mining searches [48]. STRING aggregates data and predictions stemming from a wide spectrum of cell types and environmental conditions, it aims to represent the union of all possible protein–protein links. From this union, the actual network for any given spatio-temporal snapshot of the cell can in principle be deduced by projection, for example by removing proteins known to be not expressed or not active under the conditions studied [49]. Apart from the ad hoc access through the website, STRING can be downloaded and used locally, either in the form of concise flat-files or as a complete relational database back-end.

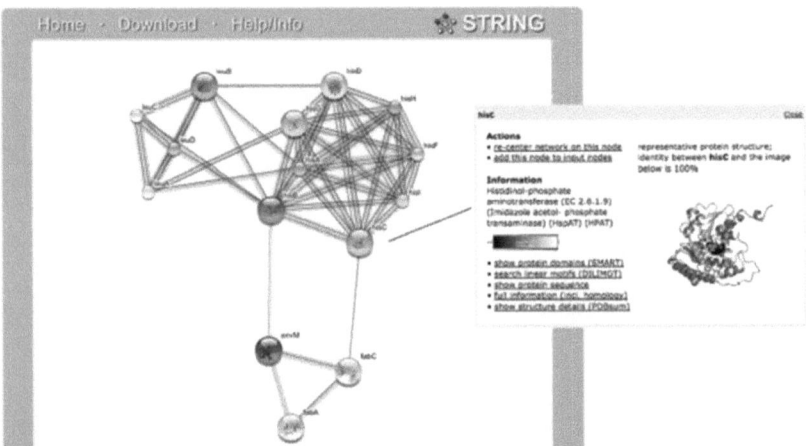

FIGURE 14 EXAMPLE PROTEIN INTERACTION NETWORK VIEW IN STRING WITH ADDITIONAL INFORMATION ABOUT SELECTED PROTEIN SHOWING IN INSET.

Figure 14 shows an example of the network view in STRING, centred on the query protein 'hisB' from *Escherichia coli*. The inset shows the annotations and options that are available for each protein, including references to other databases. Line colour indicates the type of the supporting evidence; all underlying evidence can be inspected in dedicated viewers that are accessible from the network.

2.3.5 NeAT

The network analysis tools (NeAT) provide a user-friendly web access to a collection of modular tools for the analysis of networks (graphs) and clusters (e.g. microarray clusters, functional classes, etc.) [50]. NeAT is used for analysing biological networks stored in various databases (protein interactions, regulation and metabolism) or obtained from high-throughput experiments (two-hybrid, mass-spectrometry and microarrays). The web interface interconnects the programs in predefined analysis flows, enabling the user to address a series of questions about networks of interest. The NeAT programs may be grouped in three categories: tools for manipulating graphs (graph comparison, randomization, alteration, visualization, etc.), tools for analysing clusters (or, equivalently, classes) (cluster comparison, etc.) and tools that establish the link between networks and clusters (graph clustering, graph–cluster mapping, etc.).

Figure 15 shows the summary page for a "compare-graphs" result generated on the NeAT website. The resulting graph can be displayed in yEd (http://www.yworks.com) or Cytoscape (http://www.cytoscape.org).

2 Background and related work

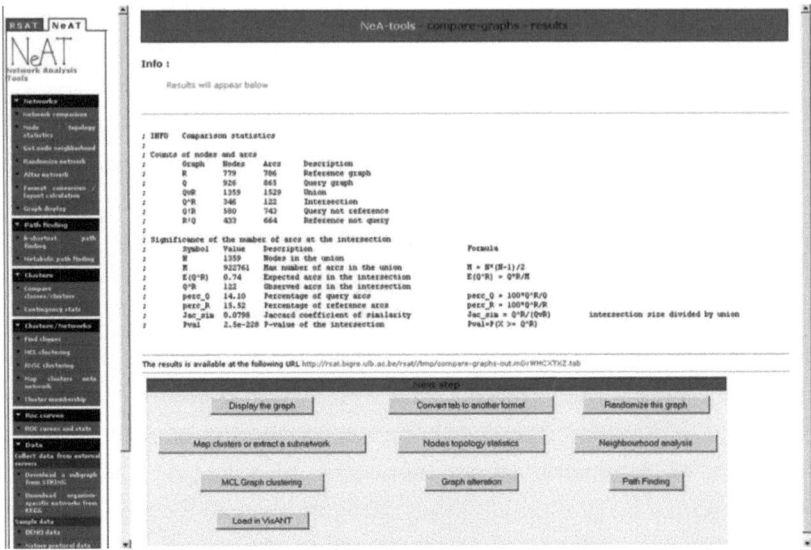

FIGURE 15 NEA-TOOLS RESULTS EXAMPLE AFTER A SUCCESSFUL "COMPARE-GRAPHS" OPERATION. DIFFERENT OPTIONS TO PROCEED ARE GIVEN AT THE BOTTOM OF THE RESULTS PAGE.

2.4 CONCLUSION

The small selection of systems presented here is enough to highlight the diversity which exists for tools for data integration in the life sciences. Some tools like Biozon or STRING focus on aspects of providing a ready integrated knowledge base to the users. On the other hand tools like SEMEDA, BNDB / BN++ or Proton provide the user with means to assemble integrated datasets on his / her own. Visual Knowledge / BioCAD or NeAT emphasise on the biological pathways and networks analysis.

Concluding from the presented systems and common practise in Systems Biology [5, 51], the representation of biological data as graphs or networks is a preferred choice. The complexity of the graphs or networks varies from tool to tool, for example NeAT works with simple node and edge lists, whereas BNDB / BN++ and SEMEDA use a semantic enriched graph model.

Graphical user interaction is realised in a variety of ways. Knowledge base focused projects like Biozon, SEMEDA or STRING tend to use a web based interface backed by a relational database. Other data integration toolkits like BNDB / BN++ or Proton offer a database driven backend with a dedicated front-end application and possible web service based access. NeAT or Visual Knowledge / BioCAD load and integrate data in an ad-hoc way as part of their analysis workflows.

All the presented tools address particular problems which they have been developed for. However, none of the presented tools was able to fulfil all our requirements as shown in the next chapter.

2 Background and related work

3 REQUIREMENTS

In this chapter the requirements for the ONDEX system with respect to both system design and functionality are being discussed. First an overview of the current situation for data integration is given, followed by describing the challenges which data integration should address. A comparison of existing approaches to data integration then leads into formulating the requirements for ONDEX.

3.1 CURRENT SITUATION

Current biological knowledge is buried in hundreds of proprietary and public life-science databases available on the World Wide Web (WWW) and millions of scientific publications [52-54]. Gaining access to this knowledge can prove difficult as each database may provide different tools to query or show the data, may differ in their structure and user interface or uses a different interpretation of biological knowledge than others. Great effort has been undertaken in order to overcome these difficulties and integrate different data sources for combining knowledge about a subject of interest. However, success of data integration is not guaranteed because of a wide range of obstacles related to access, handling and integration that exist [53].

The importance of data integration for all Life Sciences is generally recognised. Especially in the Pharmaceutical Industry [52, 55] data integration is a crucial technology for the drug discovery process [10, 56]. Effective integration of information from different biological databases and information resources is a pre-requisite for Systems Biology research [13] and has been shown to be advantageous in a wide range of use cases such as the analysis and interpretation of 'omics data [2], biomarker discovery [3] and the analysis of metabolic pathways for drug discovery [4].

3.2 CHALLENGES FOR DATA INTEGRATION

Biological knowledge such as protein interactions (Figure 16a), metabolic pathways (Figure 16b) or biological ontologies (Figure 16c) can be interpreted or understood as a network or graph. Biological databases are, however, usually implemented using table centric data structures, which do not readily allow the utilisation of graph analysis methods (**first challenge**). In this thesis a novel graph-based integration data structure will be introduced which has been developed with an emphasis on providing integrated knowledge necessary for Systems Biology. This graph-based integration data structure (see 4.1 "ONDEX integration data structure") should allow for the integration of heterogeneous data into a semantically consistent graph model and therefore support graph based analysis algorithms and visualisation.

Data integration has to face the two problems of syntactic and semantic heterogeneity [57] (**second challenge**). Syntactic heterogeneity is given by data being presented in different formats or as free text, containing spelling mistakes, wrong formatting or even missing data. Semantic heterogeneity is present in the different interpretations of data formats, symbols and names:

- Ambiguity of synonyms (exact/related), for example Na(+)/K(+)-ATPase vs. just ATPase
- Domain dependence of synonyms, for example gene names in different organisms
- Silent errors, like a typo in ENZYME Nomenclature is still valid entry (1.1.1.1 vs. 1.1.1.11)
- Unification references to other data sources can be ambiguous, for example references to multiple splicing variants of a gene assigned to a protein
- What is a Gene, what is a Protein, what is a Transcript? – biological meaning is subject to interpretation and might vary

To overcome syntactic and semantic heterogeneity in the data sources, knowledge modelling has to be adaptable for the respective domain of knowledge (see 8.2.5 "Domain knowledge") so that heterogeneous data sources can be transformed into a semantical consistent view (**third challenge**). During this process it may be necessary to identify equivalent or redundant information in the data. Novel data alignment methods (see 4.2 "Data alignment") will have to be introduced to address this need. To establish trust in the integrated data it is necessary to keep track of provenance during the whole integration process (**fourth challenge**).

3 Requirements

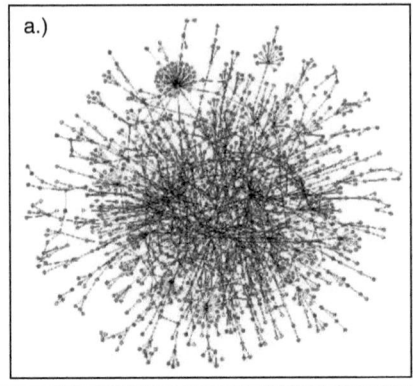

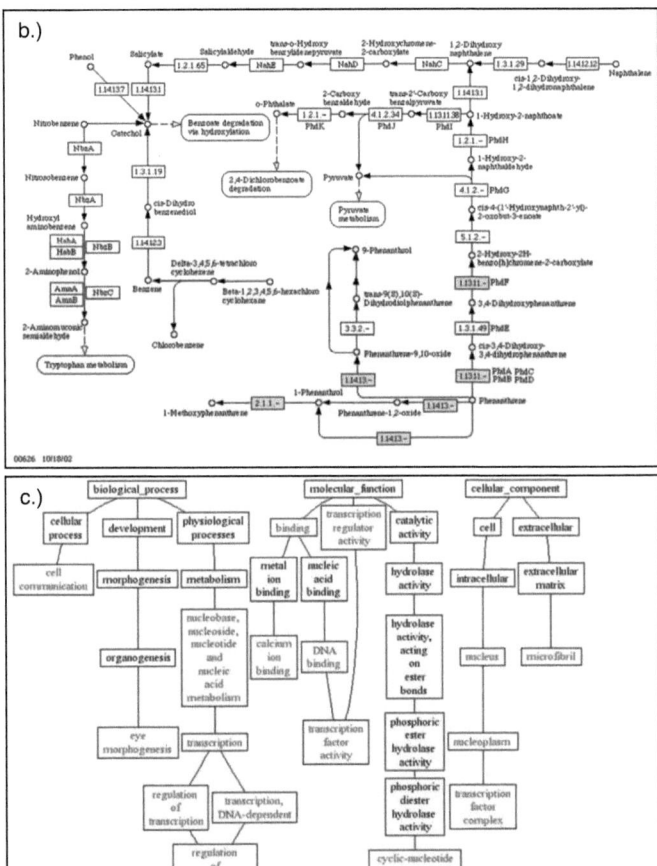

FIGURE 16 EXAMPLES OF BIOLOGICAL KNOWLEDGE AS GRAPHS: A.) PROTEIN INTERACTIONS, B.) METABOLIC PATHWAYS, C.) BIOLOGICAL ONTOLOGIES

Although this work has been mainly motivated by data from the life sciences, data integration is challenging in other data intensive sciences too. The integration methods should address this by being mostly domain independent (**fifth challenge**). An example of a different application domain for social networks will be demonstrated in the last use case (see 6.4 "Analysis of social networks"). The methods presented in this thesis have been implemented as the core of the ONDEX data integration framework [2, 5]. One key aspect of the work on ONDEX is to create a robust, usable and maintainable framework for data integration (**sixth challenge**).

TABLE 1 SUMMARISING OUTLINED CHALLENGES TO BE ADDRESSED IN THIS WORK

first challenge	representing biological data intuitively as a graph or network
second challenge	overcoming the syntactic and semantic heterogeneities between data sources
third challenge	provide a semantical consistent view on integrated information
fourth challenge	keep track of provenance during integration process
fifth challenge	domain independent approach to data integration
sixth challenge	create a robust, usable and maintainable framework for data integration

3.3 COMPARISON WITH PREVIOUS AND RELATED WORK

None of the previous presented data integration systems (see 2.2 "Previous work" and 2.3 "Survey of current data integration systems") does address all the above mentioned challenges as shown in Table 2.

3 Requirements

TABLE 2 CHALLENGES ADDRESSED BY PREVIOUS AND CURRENT WORK

	First: data intuitively as graph or network	Second: addressing syntactic & semantic conflicts	Third: semantical consistent view	Fourth: track provenance	Fifth: domain independent	Sixth: robust, usable, maintainable framework
SEMEDA	Yes	Yes	Yes	Yes	Yes	Not continued
Proton	Yes	Partially	Yes	Yes	No	Not continued
Visual Knowledge & BioCAD	Yes	No	Yes	No	No	Yes
Biozon	No	No	Yes	Yes	No	Yes
BNDB / BN++	Yes	Partially	Yes	No	Yes	Yes
STRING	Yes	No	Yes	Yes	No	Yes
NeAT	Yes	No	No	No	Yes	No

The most important aspect not or not completely addressed by previous or related work is the second challenge of addressing syntactic and semantic heterogeneities between data sources in a systematic way. Integrated knowledge base systems like STRING or Biozon use their own predefined database schema and load data from other data sources into this schema. During this process the mapping of source data to data objects in the system is hardwired and difficult to change. Overlapping or conflicting data between data sources usually does not get resolved. More complex systems like Proton, BNDB / BN++ or SEMEDA provide adapters or parsers for different data sources and let the user of the system decide which selection of data source to integrate. SEMEDA further enhances this by using lists of controlled vocabularies and information about synonyms to make more cross-references between data source than would be possible with any of the other systems. Systems like NeAT or Visual Knowledge / BioCAD rely on the data to be in the correct format involving a larger amount of manually curation and work to be done upfront.

3.4 COMPARISON OF APPROACHES TO DATA INTEGRATION

The presented approaches to data integration in Section 2.1 "Principles of data integration" are compared with each other according to the following criteria in Table 3 to be able to conclude design requirements for the ONDEX data integration framework.

TABLE 3 CRITERIA FOR THE COMPARISON OF APPROACHES TO DATA INTEGRATION

Criteria	Description	Possibilities
Up-to-date content	Are updates to the source databases visible in the integration results?	Yes/No
Local data persistence	Are data integration efforts kept persistent, for example in an underlying local database?	Yes/No
Size of content	What is the anticipated size of the complete integrated content?	Large, Medium, Small
Expected speed	What retrieval speed on average a user can expect?	Fast/Slow
Uses data model	Is an explicit data model for the integration used?	Yes/No
Complexity	How complex is the system with respect to both users and developers?	High/Low
Initial costs	What initial costs have to be anticipated for creating and deploying a respective solution?	High/Low
Maintenance costs	How difficult is it to maintain such a system and deal with changes to underlying sources?	High/Low
Expected lifetime	What is the expected lifetime of a respective solution?	Long/Short
Track provenance	How difficult is it to track the provenance of any integration result?	Hard/Easy

3 Requirements

Table 4 shows the results of the comparison followed by a short discussion.

TABLE 4 COMPARISON OF DIFFERENT APPROACHED TO DATA INTEGRATION

	Link integration and hypertext navigation	Data warehouses	View integration and mediator systems	Workflows	Mashups
Up-to-date content	Yes	No	Yes	Yes	Yes
Local data persistence	No	Yes	No	Yes/No	No
Size of content	Small	Large	Large	Medium	Small
Expected speed	Slow	Fast	Slow	Slow	Slow
Uses data model	No	Yes	Yes	Yes/No	No
Complexity	Low	High	High	High	Low
Initial costs	Low	High	High	Low	Low
Maintenance costs	Low	High	High	Low	Low
Expected lifetime	Long	Long	Long	Short	Short
Track provenance	Hard	Hard	Easy	Easy	Hard

None of the approaches used for data integration satisfies all criteria. It is always a trade-off between the complexity of creating and maintaining, the ease of use and the content and life cycle of a solution. A choice has to be made according to the intended application domain and use case. The distinction between all systems is not always as clear as it appears to be here. For example data warehouses can be controlled and interact via workflows. Link integration systems also share a high degree of similarity with Mashups.

The most significant problem common to all these data integration principles is how to address the technical and semantic heterogeneity of the diversity of life science data sources. The use of common ontologies, controlled vocabularies and naming authorities has to be encouraged to be able to unambiguously integrate data. The dawn of semantic web technologies like RDF, which uses URIs to unambiguously identify entities might help improve this situation. However, it is still a long way to go until all data and service providers might adapt such techniques and agree on unique naming and identification of biological entities, if it will ever happen. The reasons are not within the computational domain but are rooted in the political issues involved in the creation and distribution of data sets and the fractured "princely states" of the current situation [58].

3.5 REQUIREMENTS FOR ONDEX

Effective integration of information from different biological databases and information resources is a pre-requisite for Systems Biology research [13] (see Chapter 1). Yet the development of general solutions to the problem of data integration remains a significant unsolved problem in bioinformatics [57] (see 3.1 "Current situation"). At the outset of this work no single system was available that was capable of implementing the range of use cases presented later in this thesis. Furthermore data integration faces many problems and challenges as outlined in 3.2 "Challenges" with none of the previous or related systems addressing them fully (see 3.3 "Comparison with previous and related work").

The analysis of a selection of systems for data integration in the life sciences presented in Chapter 2 suggests that representing biological knowledge as graphs or semantic networks is appropriate (**first challenge**). Additionally it has become clear in Section 2.1 "Principles of data integration" that a data integration system has to be able to cope with a world in flux [9]. Hence a flexible, extensible and modular design of a data integration system is beneficial to cater for both, changes in data and changes in requirements over time.

Methods need to be developed for integrating semantically consistent information across multiple data sources, which requires that both the syntactic and semantic heterogeneity of the data sources are addressed (**second & third challenge**). It is important to keep track of provenance during the integration process (**fourth challenge**) so that trust in the data integration methods can be established and a formal evaluation of the integration results is possible. For better handling of provenance during the integration process, data alignment (see 4.2 "Data alignment", the creation of new relationships between related entities) should be used instead of technical (horizontal) integration approaches (merging of all related entities into one).

Data integration is not only a challenge in the life sciences, but for all data intensive sciences in general (**fifth challenge**). A highly configurable and generic approach for meta-data (semantics) on the graph model will enable the system to be used not only in the life sciences domain, but for multiple other domains of knowledge.

The aim of this work has been to create a general purpose environment for the integration, analysis and visualisation of complex datasets (**sixth challenge**), based on representing knowledge as a graph model. The system should incorporate aspects of a data warehouse (see 2.1 "Principles of data integration") for the following reasons (conclusions from 3.4 "Comparison of approaches to data integration"):

- control over data sources and integration process to have reliable data provenance
- efficient queries to support data mining and visualisation for detailed analysis of datasets
- there are still only few Internet data resources offering reliable Web Services

This has to be supported by easy execution of workflows to control data integration, the ability to handle very large datasets and a broad range of data exchange format support and open interfaces. Easy to use and flexible graph-based user interface(s) have to be created for the (re-)analysis of the integrated data.

In the implementation and deployment attention has to be paid to ease of use, and it would be preferable that the system ships with all required components and runs "out-of-the-box". The system should be considered as an open-source toolkit for a modestly experienced bioinformatician to develop their own integrated applications and use the provided network visualisation and analysis tools for data mining of their own datasets.

3 Requirements

4 METHODS AND PRINCIPLES

In this chapter the main methods and principles which are implemented as part of the ONDEX system are presented. In Section 4.1 the ONDEX integration data structure is defined. This data structure uses a graph based approach to facilitate representation of biological knowledge as networks, thus addressing the **first challenge**. Furthermore it provides flexibility to transform data from heterogeneous data sources into a semantical consistent view, as stated in the **third challenge**. Once data has been imported, it is possible to identify and remove redundancies in the data using mapping methods presented in Section 4.2 "Data alignment". This helps to overcome semantical heterogeneities in the data, hereby addressing the **second challenge**. The mapping methods track provenance about how exactly a relationship between entities of different data sources has been established, hereby satisfying the **fourth requirement**. The last Section 4.3 discusses issues concerning the exchange of integrated data sets and presents an application domain independent approach as proposed in the **fifth challenge**.

4.1 ONDEX INTEGRATION DATA STRUCTURE

4.1.1 MOTIVATION

The most central part in the ONDEX data integration framework is the integration data structure (see Figure 17). To be able to accommodate heterogeneous biological data from arbitrary sources (**second challenge**) requires both semantic flexibility and integrity of the integration data structure. In the context of biological data being organised as networks (for example protein interaction networks, metabolic pathways, and gene regulatory maps) a graph based integration approach has been chosen (**first challenge**). Biological entities (for example proteins, metabolites, genes) are represented by the nodes and their relationships (for example interacts, catalyses, activates) respectively by the edges in the graph. The semantics of nodes and edges in the graph are made up using a flexible type system [59] described in the next sections which enables the user to define a semantical consistent view on data (**third challenge**).

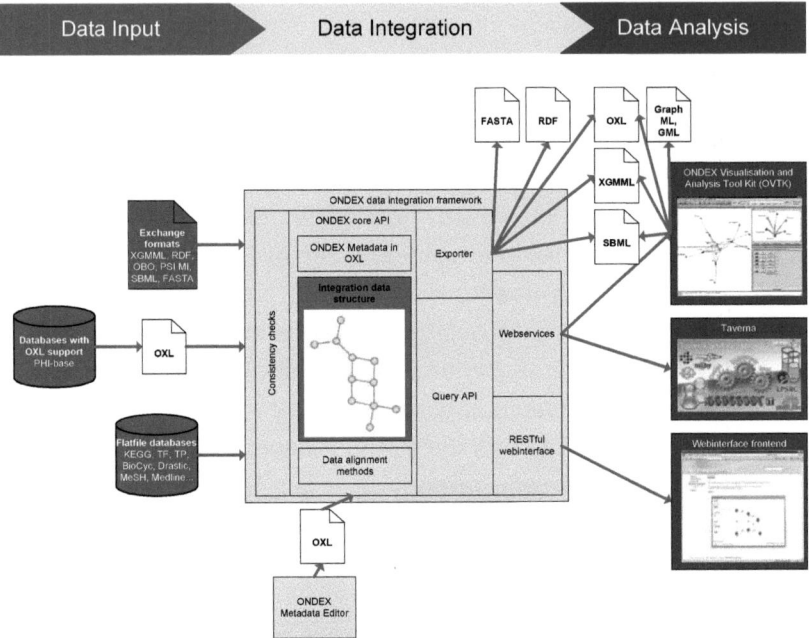

FIGURE 17 INTEGRATION DATA STRUCTURE (HIGHLIGHTED) IS THE CENTRAL COMPONENT OF THE ONDEX DATA INTEGRATION FRAMEWORK

The integration data structure incorporates features of semantic networks and conceptual graphs as presented in 8.2 "Knowledge representation" and has been inspired by the use of semantic networks for data integration in the life sciences presented in 2.3.1 "Visual Knowledge and BioCAD" as well as by the ontology driven approach presented in 2.2.1 "SEMEDA".

4.1.2 SEMANTICS IN ONDEX

There exist many definitions for the word semantics in different subject areas, for example:

- In computer science a program is described in terms of syntax and semantics. The syntax can usually be formalised using grammars, whereas the semantics of a programming language more or less defies formalisation. Semantics in computer science generally reflects the purpose of programs and functions.
- In linguistics semantics is the study of meaning of words and the relationships between different linguistic units like homonyms, synonyms, antonyms etc.
- The semantic web was defined by Tim Berners-Lee as "The Web of data with meaning in the sense that a computer program can learn enough about what the data means to process it."

All these definitions have in common that semantics refers to the meaning of something, a program, a word or some data. This interpretation of semantics has been adapted in ONDEX where it is used to convene domain specific meaning to the nodes and edges in the ONDEX integration data structure. In the case of integrated biological data the semantics on nodes describe what kind of biological entity (for example an enzyme or a metabolite) a node represents, whereas the semantics on edges describe what kind of biological relationship (for example activates enzyme or catalyses metabolite) an edge represents.

Definition:

Let O denote the ONDEX integration data structure as a graph of nodes and edges with *Concepts C(O)* := a finite, not empty, distinct set of concepts (nodes) and *Relations R(O)* ⊆ *C(O)* × *C(O)* × *(C(O)* ∪ *∅)* := a finite, not empty set of binary or ternary relations (edges)

4 Methods and principles

Example:

In case of a subset of UniProt [60] being represented as a graph in the ONDEX integration data structure (see Figure 18), the set of *Concepts C(O)* would contain all Proteins, Diseases, Enzyme Classifications (EC) and Publications, whereas the set of *Relations R(O)* would link a Protein with Diseases (if any), with an Enzyme Classification in case the Protein is an enzyme and with Publications (if any) that references this Protein.

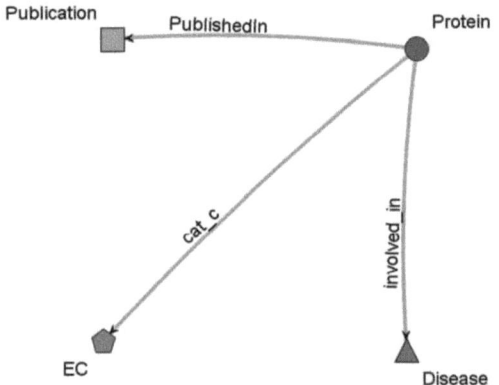

FIGURE 18 A SUBSET OF UNIPROT REPRESENTED AS A GRAPH IN THE ONDEX INTEGRATION DATA STRUCTURE WITH A PROTEIN CONCEPT (CIRCLE) CONNECTED TO CONCEPTS FOR DISEASE (TRIANGLE VIA "INVOLVED_IN" RELATION), ENYZME CLASSIFICATION (EC) (PENTAGON VIA "CAT_C" RELATION) AND PUBLICATIONS (SQUARE VIA "PUBLISHEDIN" RELATION).

4.1.3 SEMANTICS OF NODES

A hierarchy of classes is used to model knowledge about the entities in the graph. These classes define the type of a concept (node) in the graph similar to conceptual graphs (see 8.2.2 "Conceptual graphs"). The hierarchy is constituted by subsumption relationships and therefore defines a taxonomy (see 8.2.7 "Hierarchy and taxonomies"), i.e. one class can be a specialisation of another class and therefore forms a rooted tree. A common root for this hierarchy is the class "Thing" representing any kind of object that exists or can exist similar to the ontology used in 2.2.1 "SEMEDA". Exceptions are allowed for classes describing concepts or ideas, for example a word or thought. Such classes are usually not rooted. Different roots may exist to distinguish between different domains of knowledge, for example biological versus social knowledge domain. If more than one such tree representing a hierarchy is present it is called a forest. All hierarchies together make up the ontology of possible types of concepts in the integration data structure.

Definition:
> Concept Classes $CC(O) \coloneqq$ a taxonomy of concept classes, with the totally defined function cc that assigns exactly one concept class to each concept $cc: C(O) \rightarrow CC(O)$

Example:
> The Sequence Ontology [61] (Figure 19) can readily be used for constituting the concept classes in the graph as an example. One concept could have the concept class *Protein Coding Gene*, whereas another concept could have the concept class *Region* and be referred to via a relation (edge) from the former *Protein Coding Gene*.

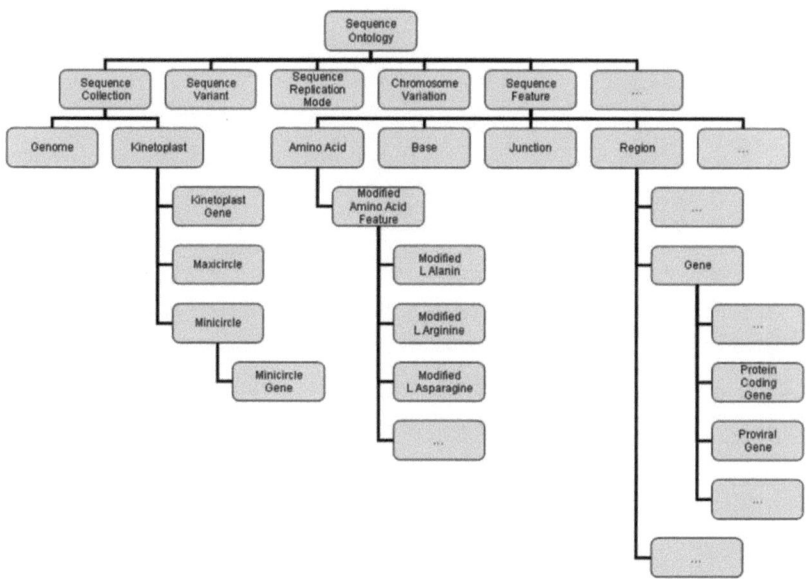

FIGURE 19 EXCERPT FROM THE SEQUENCE ONTOLOGY SHOWING HIERARCHY OF CLASSES ROOTED IN DOMAIN SPECIFYING NODE "SEQUENCE ONTOLOGY"

Figure 19 is an excerpt from the Sequence Ontology [61] showing the hierarchy of classes to describe sequence related properties. In this example a *Protein Coding Gene* "is specialisation of" a *Gene* "is specialisation of" a *Region* "is specialisation of" a *Sequence Feature*, which itself is part of the *Sequence Ontology* domain of knowledge (see 8.2.5 "Domain knowledge").

An entity in the graph belongs to exactly one class in a given hierarchy, the concept class. Therefore it is always possible to associate a concept with a unique domain of knowledge as each domain has its own hierarchy. This facilitates the representation of multiple domains of knowledge (e.g. Sequence Ontology and Gene Ontology) at the same time within the same integration data structure.

4.1.4 SEMANTICS OF EDGES

Edges in the integration data structure represent conceptual relations in contrast to conceptual graphs (see 8.2.2 "Conceptual graphs") where the conceptual relation is represented as a node. A relation carries a semantic meaning similar to edges in semantic networks (see 8.2.1 "Semantic network") and therefore has a type assigned. A hierarchy of types is used to model knowledge about the relationships in the graph. The hierarchy is constituted by a subsumption relationship similar to semantics on nodes, i.e. one type can be a specialisation of another type and therefore form a rooted tree. The common root for such a hierarchy is the type "related" representing any kind of relationship that exists or can exist between two entities. In contrast to semantics on nodes, no un-rooted types are allowed and all hierarchies should be rooted at the common root "related". Relationships are directed, yet relation types can explicitly state the inverse relationship. Relationships can span across multiple domains of knowledge (see 8.2.5 "Domain knowledge"), for example the "equivalence" relationship may be used to relate two entities from separate knowledge domains to each other.

TABLE 5 OBO RELATION ONTOLOGY SHOWING PROPERTIES OF DIFFERENT RELATION TYPES

name	transitive	symmetric	reflexive	anti-symmetric
is_a	+		+	+
part_of	+		+	+
integral_part_of	+		+	+
proper_part_of	+			
located_in	+		+	
contained_in				
adjacent_to				
transformation_of	+			
derives_from	+			
preceded_by	+			
has_participant				
has_agent				
instance_of				

Table 5 shows the OBO Relation Ontology [62], which is an ontology of core relations for use by OBO Foundry ontologies. The *is_a* relationship is considered axiomatic by the OBO file format specification, and by OWL [63]. There is a direct translation of this axiomatic relationship to the subsumption relationship used in the hierarchy of concept classes in the integration data structure. For example given the Sequence Ontology (see Figure 19) a concept belonging to the class *Protein Coding Gene* "is a" instance of *Gene* given implicitly by the class hierarchy subsumption relationship.

Definition:

Relation Types $RT(O)$:= a taxonomy of relation types, with the totally defined function rt that assigns exactly one relation type to each relation $rt: R(O) \rightarrow RT(O)$

Example:

The OBO Relation Ontology (see Figure 20) can readily be used to describe the relationship between concepts. For example a *Protein Coding Gene* (concept) can be *located_in* (relation) a particular *Region* (concept). Here concept classes for concepts of *Protein Coding Gene* and *Region* have been taken from the previous example of Sequence Ontology (see Figure 19).

The OBO Relation Ontology itself is a hierarchy constituted by a subsumption relationship and rooted at "relationship" ("A relationship between two classes (terms). Relationships between classes are expressed in terms of relations on underlying instances. [64]"). This root can be directly translated into the common root "related" of the hierarchy of types facilitating the semantics on edges in the graph. Furthermore all relationships within the OBO Relation Ontology (see Figure 20) can be captured using the subsumption relationship.

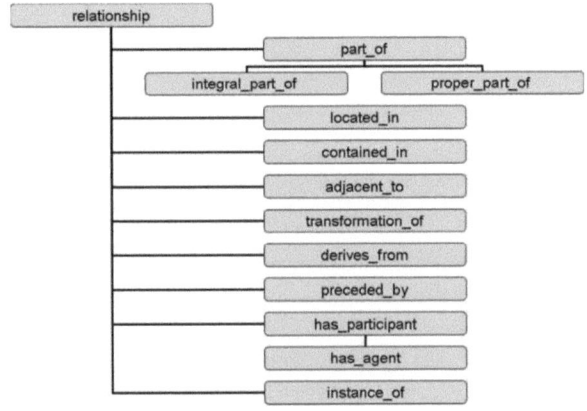

FIGURE 20 OBO RELATION ONTOLOGY SHOWING DIFFERENT RELATION TYPES AND THEIR HIERARCHY ROOTED IN THE MOST GENERAL TYPE "RELATIONSHIP"

The semantics on the edges of the graph are motivated by the OBO Relation Ontology. Therefore any relation type associated with the edges in the graph has the additional four attributes transitive, symmetric, reflexive and anti-symmetric (see 8.1.2 "Properties on relations"). Although the OBO Relation Ontology is an important building block, it alone is not able to capture all the relationships of all domains of knowledge, for example that an enzyme catalyses a reaction. Thus domain dependent extensions to the hierarchy of relation types will be made in the particular case and all together will constitute the ontology of possible types of relations in the integration data structure.

4.1.5 PROVENANCE

Provenance on nodes and edges of the integration data structure is realised using a two-pronged approach based on the originating data source and evidence type.

Firstly, the originating data source or controlled vocabulary (CV, see 8.2.8 "Controlled vocabulary") is added as an attribute to the nodes similar to the use of CV in 2.2.1 "SEMEDA". Each node always belongs to exactly one data source, even if such a data source is the result of the union of different data sources. Edges in the graph have implicit provenance about the originating data source derived through reasoning over the connected nodes. For entities in the graph which are not present in any existing data source, i.e. derived or manually created concepts, a special convention is used, assigning the CV "virtual ontology" (VO) as the data source.

Definition:
> *Controlled Vocabularies* $CV(O) \coloneqq$ a finite, not empty set of controlled vocabulary identifiers, with the totally defined function cv that assigns exactly one controlled vocabulary identifier to each concept $cv: C(O) \rightarrow CV(O)$

Example:
> Many controlled vocabularies or data sources exist which can be integrated in ONDEX, for example UniProt [60], KEGG [65], BioCyc [66] and so on. Each of these data sources has its own unique identifier in the ONDEX integration data structure. In the case of concepts representing entries from UniProt, these concepts would carry the UniProt unique identifier.

Secondly, one or more evidence types or codes are assigned to the entities in the integration data structure. Evidence types reflect how the particular entity in the graph has been created and / or modified. Different data integration methods will use different evidence types. The most common case is "imported from data source" (IMPD), which simply states that information about an entity in the graph is the same as in the data source it was imported from. Evidence types are closely coupled with the data integration process and reflect the application domain, i.e. different evidence types could be used in different application domains.

Definition:
> Evidence Types $E(O) \coloneqq$ a finite, not empty set of evidence types, with the surjective function e which assigns evidences to every concept and every relation $e: C(O) \cup R(O) \rightarrow \{(e_1 \times ... \times e_n) | e_j \in E(O)\}$

Example:
> Evidence types identify how entities in the integration data structure have been created. In the case of manually adding elements to the integration data structure, which are not from any known data source, an appropriate evidence type would be "manually created" (M). Most data alignment methods (see 4.2 "Data alignment") assign a unique evidence type to relations they compute, for example "synonym mapping" (SYN) or "StructAlign" (STRUCT) would be appropriate evidence types in these cases.

4.1.6 REFERENCES AND SYNONYMS

Data sources are commonly linked with each other using a particular reference scheme from data source A into data source B. Such cross references are captured as a special attribute on nodes in the integration data structure called an accession. An accession is a pair composed from the reference and the data source the reference is pointing to. For example, a node representing an entry in data source A would have an accession containing the particular reference and a pointer to data source B. If no such cross reference exists then the accession attribute is empty. Nodes can have an arbitrary number of accession attributes.

The reference schemes between data sources can be of different quality, i.e. they may represent one-to-one relationships or one-to-many relationships. One-to-one relationships imply identity of the entries between the data sources. One-to-many relationships are usually the case when one data source contains more specialised information than the other. In this case a more general term could be

4 Methods and principles

referenced from many specialised terms. Such accessions will be flagged as ambiguous.

Definition:
>   *Concept Accessions* $CA(O) := $ a finite set of concept accessions, with the function ca which assigns concept accessions to a concept $ca: C(O) \rightarrow \{(ca_1 \times ... \times ca_n) | ca_j \in CA(O)\}$

Example:
>   The UniProt [60] protein *P00330* (ADH1_YEAST) has cross database references into other databases, for example PDB [67]. In PDB this protein has the identifier *2HCY*. Therefore in the integration data structure *P00330* would have a concept accession ("PDB","2HCY") which is not ambiguous as both databases speak about the same protein concept. On the other hand the cross database reference to SGD [68] with identifier *S000005446* would be ambiguous, as SGD focuses on open reading frames (ORF) belonging to the protein and not the protein concept itself.

Synonyms are human readable names assigned to the nodes of the graph. The list of synonyms on a node can be empty. Synonyms are an additional way of communicating that more than one name is used to refer to the same entity or concept. Multiple synonyms are a problem throughout biology, where for example there are different names for the same gene in different organism. There are also multiple ways of naming most chemical compounds. Synonyms can be grouped into exact synonyms and related synonyms. Exact synonyms describe only one particular concept, whereas related synonyms may describe a group or set of concepts, for example alcohol dehydrogenase is an exact synonym and oxidoreductase is a related synonym for entry 1.1.1.1 of the Enzyme Nomenclature [69].

Definition:
> *Concept Names* $CN(O) := $ a finite set of concept names, with the function cn which assigns concept names to a concept
> $cn: C(O) \rightarrow \{(cn_1 \times ... \times cn_n) | cn_j \in CN(O)\}$

Example:
> The UniProt [60] protein *P00330* (ADH1_YEAST) has the recommend name from UniProt *Alcohol dehydrogenase 1*, which would be recognized as an exact synonym (concept name) for the concept *P00330* in the integration data structure. The alternative names *Alcohol dehydrogenase I* and *YADH-1* on the other hand would not be marked as exact synonyms.

4.1.7 GENERALISED DATA STRUCTURE

Storing heterogeneous information from arbitrary domains of knowledge (see 8.2.5 "Domain knowledge") requires a flexible data structure. The Generalised Data Structure (GDS) consists of an attribute name and a value.

An attribute name expresses domain specific knowledge about the value. This can include information about the unit of the value, for example kilograms, and the syntax of the value, for example being a number. Attribute names can be arranged as a hierarchy using the subsumption relationship. Attribute names do not have to be rooted. They can belong to different domains of knowledge, for example an attribute name representing molecular weight or one representing the sex of an animal might both be assigned to one concept. Or they can be domain independent like capturing a certain time or duration.

The value is treated as a complex object as defined by the attribute name. A value should not be empty. The non-existence of a value on a particular entity in the graph is implicitly given. Nodes and edges in the graph can have arbitrary lists of GDS.

Definition:
> *Generalised Data Structure GDS* $P(O) :=$ a finite set of additional properties (GDS), with the function p which assigns properties (GDS) to a concept or a relation $p: C(O) \cup R(O) \rightarrow P(O)^*$

Example:
> The UniProt [60] protein *P00330* (ADH1_YEAST) has an amino acid sequence of length 348 as a property. Both the length of the sequence and the sequence itself can be captured using GDS. For the length of the sequence an attribute name (LENGTH) specifying a number data type is required, whereas the sequence itself would have an attribute name (SEQ) with data type character sequence (String). The resulting GDS on concept *P00330* as pairs of attribute name and value would be (LENGTH, 348) and (SEQ, "MSIPETQKG…").

4.1.8 CONTEXT

In conceptual graphs, contexts (see 8.2.4 "Contexts") are represented by concept boxes that contain nested graphs that describe the referent of the concept. Similar to conceptual graphs context in the integration data structure is used to describe nested graphs. For example a nested graph could be describing a semantical consistent view on one particular biological pathway or a certain cellular location. Rather than having all concepts and relations of the nested graph as referents of the concept representing the context, each context declaring concept is assigned to the concepts and relations of the nested graph itself. This inverse context structure abolishes the requirement for concepts to represent context directly, i.e. have the nested graph as referents like in conceptual graphs. Therefore any concept contained in the graph can be blessed to constitute the context of a nested graph.

Definition:
> Context is the function co which assigns particular concepts as context to a concept or relation $co: C(O) \cup R(O) \to C(O)^*$

Example:
> The KEGG [65] database contains biological pathways, for example path:ath00010. This pathway could be represented by a concept *PATH:ATH00010* in the integration data structure. Elements of path:ath00010 such as enzymes and metabolites would also be represented as concepts, for example the alcohol dehydrogenase from *Arabidopsis thaliana* *AT1G22430*. Then the concept *PATH:ATH00010* can be added to the context list on concept *AT1G22430*. Thus implicitly defining that concept *AT1G22430* is part of the special nested graph tagged or labelled with concept *PATH:ATH00010*.

4.1.9 DEFINITION ONDEX INTEGRATION DATA STRUCTURE

This section combines all definitions from the previous sections into one coherent definition for the integration data structure.

The ONDEX integration data structure O is defined as a tuple
$O := (C, R, CA, CN, CV, CC, RT, E, P, ca, cn, cv, cc, rt, e, p, co, id)$
consisting of:
- a finite, not empty, distinct set of concepts $C(O)$
- a finite, not empty set of relations $R(O) \subseteq C(O) \times C(O) \times (C(O) \cup \emptyset)$
- a finite set of concept accessions $CA(O)$
- a finite set of concept names $CN(O)$
- a finite, not empty set of controlled vocabulary identifiers $CV(O)$
- a taxonomy of concept classes $CC(O)$
- a taxonomy of relation types $RT(O)$
- a finite, not empty set of evidence types $E(O)$
- a finite set of additional properties (GDS) $P(O)$

- the function ca which assigns concept accessions to concepts
    - $ca: C(O) \rightarrow \{(ca_1 \times ... \times ca_n) | ca_j \in CA(O)\}$
- the function cn which assigns concept names to concepts
    - $cn: C(O) \rightarrow \{(cn_1 \times ... \times cn_n) | cn_j \in CN(O)\}$
- the totally defined functions cv, cc, rt that assign controlled vocabulary identifiers, concept classes and relation types to concepts or relations
    - $cv: C(O) \rightarrow CV(O)$
    - $cc: C(O) \rightarrow CC(O)$
    - $rt: R(O) \rightarrow RT(O)$
- the surjective function e which assigns evidences to every concept and every relation
    - $e: C(O) \cup R(O) \rightarrow \{(e_1 \times ... \times e_n) | e_j \in E(O)\}$
- the function p which assigns properties (GDS) to concepts and relations
    - $p: C(O) \cup R(O) \rightarrow P(O)^*$
- the function co which assigns a context to concepts and relations
    - $co: C(O) \cup R(O) \rightarrow C(O)^*$
- the bijective function id which assigns a unique number identifier to every concept and every relation
    - $id: C(O) \cup R(O) \rightarrow \mathbb{N}$

4.1.10 Discussion

The ONDEX integration data structure reflects notions of both semantic networks (see 8.2.1 "Semantic network") and conceptual graphs (see 8.2.2 "Conceptual graphs"). Nodes represent concepts with types, called concept classes that form a taxonomy. Edges represent conceptual relations associated with relation types arranged in a taxonomy. Arbitrary properties of concepts and relations can be assigned using GDS. The notion of conceptual graphs has been extended via certain attributes with defined semantics on nodes and edges and the use of ternary relations. These attributes on concepts and relations are evidence type, unique identifier and context. Additionally concepts have concept accessions, concept names and data source (CV). A conceptual relation is expressed as an arc in the integration data structure, which is more similar to semantic networks. The difference in representing context compared to conceptual graphs provides additional flexibility as the need for dedicated "context nodes" like in conceptual graphs is abolished.

4.2 DATA ALIGNMENT

Parts of this chapter have been published in the Proceedings of the 6th International Workshop on Data Integration in the Life Sciences in 2009, see [6].

4.2.1 MOTIVATION

Software designed to integrate data for life sciences applications has to address two classes of problem (see **second challenge** in 3.2 "Challenges for data integration"). It must provide a general solution to the technical (syntactic) heterogeneity, which arises from the different data formats, access methods and protocols used by different databases. More significantly, it must address the semantic heterogeneities arising from a number of sources in life science databases. The most challenging source of semantic heterogeneity comes from the diversity and inconsistency among naming conventions for genes, gene functions, biological processes and structures among different species (or even within species). In recent years, significant progress in documenting the semantic equivalence of terms used in the naming of biological concepts and parts has been made in the development of a range of biological ontology databases which are coordinated under the umbrella of organisations such as the Open Biomedical Ontologies Foundry (http://www.obofoundry.org). However the majority of biological terms still remains uncharacterised and therefore requires automated methods to define equivalence relationships between them.

The integration of data in ONDEX generally follows three conceptual stages as illustrated in Figure 21: (1) normalising into a integration data structure in order to overcome predominantly technical heterogeneities between data exchange formats; (2) identifying equivalent and related entities among the imported data sources to overcome semantic heterogeneities at the entry level, and (3) the data analysis, information filtering and knowledge extraction.

4 Methods and principles

In order to make the ONDEX system as extensible as possible, the second stage (red highlighted in Figure 21) has been separated both conceptually and practically. The motivations for doing this are to: preserve original relationships and metadata from the original data source; make this semantic integration step easily extensible with new methods; implement multiple methods for recognising equivalent data concepts to enhance the quality of data integration and support reasoning methods that make use of the information generated in this step to improve the quality of data integration (**third & fourth challenge**).

The hypothesis here is that multiple methods for semantic data integration are necessary because of ambiguities and inconsistencies in the source data that will require different treatment depending on the source databases. In many cases, exact linking between concepts through unique names will not always be possible and therefore mappings will need to be made using inexact methods. Unless these inexact methods can be used reliably, the quality of the integrated data will be degraded.

To calibrate the presented data integration methods with well-structured data, the mapping of equivalent elements from the ontologies and nomenclatures extracted from the ENZYME [70] and GO [71] databases is used. To evaluate mapping methods in a more challenging integration task, the creation of an integrated dataset from two important biological pathway resources, the KEGG [72] and AraCyc [73] databases, is presented.

4 Methods and principles

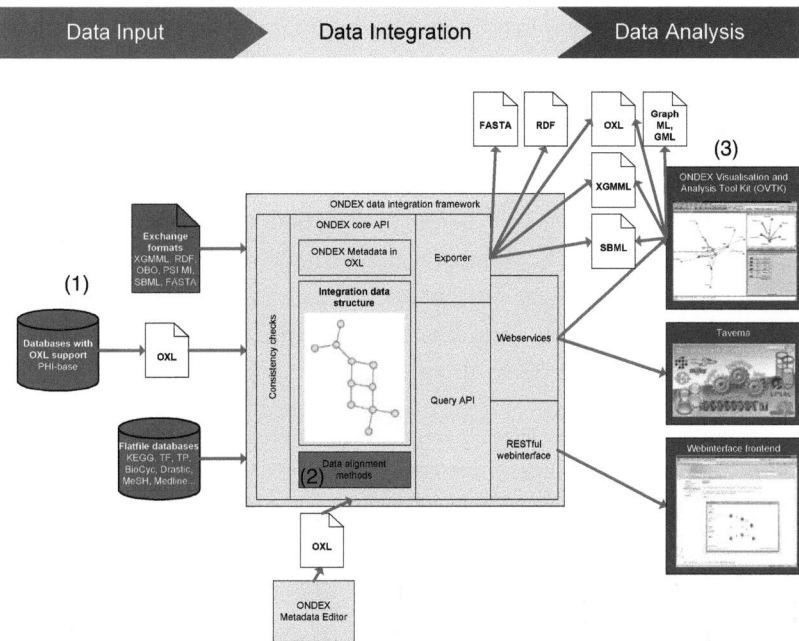

FIGURE 21 DATA INTEGRATION IN ONDEX CONSISTS OF 3 STEPS. 1) IMPORT AND CONVERSION OF DATA SOURCES INTO THE INTEGRATION DATA STRUCTURE OF ONDEX (DATA INPUT LEFT), 2) LINKING OF EQUIVALENT OR RELATED ENTITIES OF THE DIFFERENT DATA SOURCES (DATA INTEGRATION MIDDLE), 3) KNOWLEDGE EXTRACTION IN THE GRAPH BASED ANALYSIS COMPONENT (DATA ANALYSIS RIGHT).

4.2.2 METHODS

4.2.2.1 DATA IMPORT AND EXPORT

Following Figure 21, the first data integration step loads and indexes data from different sources. ONDEX provides several options for loading data into the internal data warehouse and a range of parsers have been written for commonly used data sources and exchange formats (see 9.1 "List of data formats supported by ONDEX"). In addition users can convert their data into an ONDEX specific XML or RDF dialect for which generic parsers (see 4.3 "Exchanging integrated data") are provided.

4 Methods and principles

The role of all parsers is to load data from different data sources into the integration data structure used in the ONDEX data integration framework. In simple terms, this integration data structure can be seen as a graph, in which concepts are the nodes and relations are the edges. By analogy with the use of ontologies for knowledge representation in computer science, concepts are used to represent real-world objects [74]. Relations are used to represent the different ways in which concepts are connected to each other. Furthermore, concepts and relations may have additional properties and optional characteristics attached to them (see 4.1 "ONDEX integration data structure").

During the import process, names for concepts are lexicographically normalised by replacing non-alphanumeric characters with white spaces so that only numbers and letters are kept in the name. In addition, consistency checks are performed to identify, for example, empty or malformed concept names.

4.2.2.2 DATA INTEGRATION METHODS AND ALGORITHMS

The second data integration step (following Figure 21) links equivalent and related concepts and therefore creates relations between concepts from different data sources. Different combinations of mapping methods can be used to create links between equivalent or related concepts. Rather than merging elements that are found to be equivalent, the mapping methods create a new equivalence relation between such concepts, which is known as data alignment. Each mapping method can be made to create a score value reflecting the belief in a particular mapping and information about the parameters used. These scores are assigned as edge weights to the graph and form the foundation for the statistical analysis presented later. Additionally information on edges enables the user to track evidence for why two concepts were mapped by a particular mapping method.

Several constraints must be fulfilled before a mapping method creates a new link between two concepts. Under the assumption that the integrated data sources already contain all appropriate links between their own entries, new links are only created between different data sources (CVs). Biological databases often provide an NCBI taxonomy identifier for species information associated with their entries. If such identifiers are found in the graph, the mapping method ensures in most cases, that relations are only created within the same species. In addition to species restriction a mapping method takes the concept class of a concept into account. Only equal concept classes or specialisations of a concept class are considered to be included in a mapping pair.

ACCESSION BASED MAPPING: Most of the well-structured and managed public repositories of life science data use accession coding systems to uniquely identify individual database entries. These codes are persistent over database versions. Cross references between databases of obviously related data (for example protein and DNA sequences) can generally be found using accession codes and these can be easily exploited to link related concepts. Such concept accessions may not always present a one-to-one relationship between entries of different databases. For example, a GenBank accession found in the KEGG database is only unique for the coding region on the genome and not for the expressed proteins, which may exist in multiple splice variants. References presenting one-to-many relationships are call ambiguous. Concept accessions are indexed for better performance during information retrieval. Accession based mapping by default uses only non-ambiguous concept accessions to create links between equivalent concepts, i.e. concepts that share the same references to other databases in a one-to-one relationship. This behaviour can be changed using a parameter.

Pseudo-code:

Let O denote the ONDEX integration data structure (see 4.1 "ONDEX integration data structure") consisting of a set of concepts $C(O)$ and a set of relations $R(O) \subseteq C(O) \times C(O) \times (C(O) \cup \emptyset)$. Every concept $c \in C(O)$ has a concept class $cc(c) \in CC(O)$, a controlled vocabulary $cv \in CV(O)$ and a list of concept accessions $ca(c) = \{(ca_1 \times ... \times ca_n) | ca_j \in CA(O)\}$. Each concept accession $ca \in CA(O)$ is a triple $ca = (cv, acc, ambiguous)$, where cv is the data source identifier from which the accession code acc is derived and ambiguous is either true or false. The bijective function id assigns a consecutive number $n \in \mathbb{N}$ to concepts and relations in O separately starting with 1.

```
ignoreAmbiguity ← true or false (default)
function AccessionBasedMapping(O, ignoreAmbiguity) {
    for all i ∈ [1..|C(O)|] do
        for all j ∈ [i..|C(O)|] do
            if ∃x ∈ ca(cᵢ) ∧ x ∈ ca(cⱼ) ∧ (¬x.ambiguous ∨ ignoreAmbiguity) do
                if cv(cᵢ) ≠ cv(cⱼ) ∧ cc(cᵢ) = cc(cⱼ) do
                    O.createRelation(cᵢ, cⱼ)
}
```

Runtime analysis:

Assuming that the test if $ca(c_i)$ and $ca(c_j)$ have at least one concept accession in common takes linear time with respect to the length of the list, for example by using hashing strategies or ordered lists, and the average number of concept accessions on concepts is μ_{ca} then the total runtime of Accession based mapping is: $T(n) = \frac{1}{2}(n^2 + n) * \mu_{ca} \in O(n^2)$ where n is the number of concepts.

SYNONYM MAPPING: Entries in biological data sources often have one or more human-readable names, for example gene names. Depending on the data source, some of these names will be exact synonyms such as the chemical name of an enzyme; others only related synonyms such as a term for enzymatic function. Exact synonyms are flagged as preferred concept names during the import process. Related synonyms are added to concepts as additional concept names. Concept names are pre-processed to strip all non-letter characters and stem special word cases before inserting them into the full-text index. Concept names are indexed for better performance and potentially fuzzy searches during information retrieval using the Apache Lucene (http://lucene.apache.org/) full-text indexing system. Fuzzy searches as supported by Lucene can be useful to overcome spelling mistakes, for example PKM2 might be written as PK-M2 [75]. The default method for synonym mapping creates a link between two concepts if two or more preferred concept names are matching (bidirectional best hits) to be able to cope with ambiguity of names. As a simple example of such ambiguity, the term "mouse" shows that consideration of only one synonym is usually not enough for the disambiguation of the word, i.e. "mouse" can mean computer mouse or the rodent *Mus musculus*. The threshold for the number of synonyms to be considered a match and an option to use only exact synonyms are parameters in the synonym mapping method.

Pseudo-code:

Let O denote the ONDEX integration data structure (see 4.1 "ONDEX integration data structure") consisting of a set of concepts $C(O)$ and a set of relations $R(O) \subseteq C(O) \times C(O) \times (C(O) \cup \emptyset)$. Every concept $c \in C(O)$ has a concept class $cc(c) \in CC(O)$, a controlled vocabulary $cv \in CV(O)$ and a list of concept names $cn(c) = \{(cn_1 \times ... \times cn_n) | cn_j \in CN(O)\}$. Each concept name $cn \in CN(O)$ is a tuple $cn = $ (name,preferred), where name is the actual name of the concept and preferred is either true or false. The bijective function id assigns a consecutive number $n \in \mathbb{N}$ to concepts and relations in O separately starting with 1.

```
num ← 1..N (default: 2)
exact ← true (default) or false
function SynonymMapping(O, num, exact) {
  for all i ∈ [1..|C(O)|] do
    for all j ∈ [i..|C(O)|] do
      if |cn(cᵢ) ∩ cn(cⱼ)| ≥ num ∧ (∃x ∈ cn(cᵢ) ∩ cn(cⱼ)|x.preferred ∨ ¬exact) do
        if cv(cᵢ) ≠ cv(cⱼ) ∧ cc(cᵢ) = cc(cⱼ) do
          O.createRelation(cᵢ, cⱼ)
}
```

Runtime analysis:

Assuming that the intersection of $cn(c_i)$ and $cn(c_j)$ can be found in linear time to the size of the lists by using ordered lists and the average number of concept names per concept is μ_{cn}, then the total runtime of Synonym mapping is: $T(n) = \frac{1}{2}(n^2 + n) * \mu_{cn} \in O(n^2)$ with n is the number of concepts.

STRUCTALIGN MAPPING: In some cases, two or more synonyms for a concept are not available in the integrated data. To disambiguate the meaning of a synonym shared by two concepts, the *StructAlign* mapping algorithm considers the graph neighbourhood of such concepts. A breadth-first search of a given depth (≥1) starting at each of the two concepts under consideration yields the respective reachability list for each concept. *StructAlign* processes these reachability lists and searches for synonym matches of concepts at each depth of the graph neighbourhood. If at any depth one or more pairs of concepts which share synonyms are found, *StructAlign* creates a link between the two concepts under consideration.

Pseudo-code:

Let O denote the integration data structure (see 4.1 "ONDEX integration data structure") consisting of a set of concepts $C(O)$ and a set of relations $R(O) \subseteq C(O) \times C(O) \times (C(O) \cup \emptyset)$. Every concept $c \in C(O)$ has two additional attributes assigned: a.) a concept class $cc(c)$ characterising the type of real world entity represented by the concept (for example a gene), b.) a controlled vocabulary $cv(c)$ stating the data source (for example KEGG) the concept was extracted from. Every relation $r \in R(O)$ is a triple $r = (f, t, q)$ with f the "from"-concept, t the "to"-concept and q the "qualifier"-concept (can be empty) of the relation. To improve performance the algorithm is using indexing structures for concept names and a unique identifier returned by the bijective function id which assigns a consecutive number $n \in \mathbb{N}$ to concepts and relations in O separately starting with 1.

```
index ← searchable index of concept names for concepts
cutoff ← maximal depth of graph neighbourhood
function StructAlign(O, index, cutoff) {
  matchesMap ← new map of concepts to concept sets
  // search for concept name matches
  for all c ∈ C(O) do
    for all n ∈ cn(c) | n.preferred do
      matches ← index.search(n.name)
      for all c' ∈ matches with cc(c) = cc(c') ∧ cv(c) ≠ cv(c') do
        matchesMap[c].add(c')
  connectedMap ← new map of concepts to concept sets
  // calculate direct neighbourhood
  for all r ∈ R(O) with r = (f, t, _) do
    if cv(f) = cv(t) ∧ f ≠ t do
      connectedMap[f].add(t)
      connectedMap[t].add(f)
  reachabilityMap ← clone(connectedMap)
  // modified breadth first search with depth cutoff
  for all i ∈ [1..cutoff] do
    for all (x, (y₁ ... yₙ)) ∈ reachabilityMap do
      for all j ∈ [1..n] do
        reachabilityMap[x].addAll(connectedMap[yⱼ])
  // look at neighbourhood of bidirectional matches
  for all (a, (b₁ ... bₙ)), (bᵢ, (c₁ ... cₘ)) ∈ matchesMap|a ∈ (c₁ ... cₘ) do
    na ← reachabilityMap[a]
    nb ← reachabilityMap[bᵢ]
    for all x ∈ na do
      if ∃y ∈ matchesMap[x]|y ∈ nb do
        O.createRelation(a, bᵢ)
}
```

Runtime analysis:

Assuming the search for a concept name in the list of concept names takes logarithmic time with respect to the length of the list (for example using a self-balancing binary search tree [76]) and operations to manipulate maps and sets take constant time using hashing strategies, the runtime analysis is: Let c be the number of concepts, μ_{cn} the average number of concept names associated with a concept, r be the number of relations, μ_r the average number of relations per concept in the integration data structure and Δ a time constant for operations on maps and sets. The worst case runtime of the StructAlign algorithm is then:

1.) Search for concept name matches
$$T_1(c,r) = c * \mu_{cn} * \log(c * \mu_{cn}) * c * \Delta$$
2.) Calculation of direct neighbourhood
$$T_2(c,r) = r * 2 * \Delta$$
3.) Modified breath first search with depth cut-off
$$T_3(c,r) = \text{cutoff} * c * \mu_r * \Delta$$
4.) Finding bidirectional matches in neighbourhood, $\log(c)$ search time for $\exists y$.
$$T_4(c,r) = c^2 * c * \Delta$$

$T(c,r,) = T_1 + T_2 + T_3 + T_4$
$T(c,r) = c * \mu_{cn} * \log(c * \mu_{cn}) * c * \Delta + r * 2 * \Delta + \text{cutoff} * c * \mu_r * \Delta + c^2 * c * \Delta$

Within a fully connected graph the number of relations is $r = c * (c-1)/2$ and $\mu_r = c - 1$.
$T(c) = (c * \mu_{cn} * \log(c * \mu_{cn}) * c + c * (c-1) + \text{cutoff} * c * (c-1) + c^2 * c) * \Delta$
$T(c) = (c^2 * \mu_{cn} * \log(c * \mu_{cn}) + (1 + \text{cutoff}) * c * (c-1) + c^3) * \Delta$
$T(c) \in O(c^3)$

Here the average number of concept names per concept is $\mu_{cn} \ll c$. Hence the algorithm has a worst case runtime of $O(c^3)$. Although the expected runtime on sparse graphs is $O(c^2)$ as the number of neighbours reachable for a certain depth in a sparse graph is much smaller than the number of total concepts in the graph.

Worked example for StructAlign:

Figure 22 shows a simple example graph of metabolites (circles) and enzymes (rectangles) originating from two data sources DB1 (blue) and DB2 (purple). All concepts except for concept 2 have two synonyms (preferred listed first). The "consumes" relation (red) is present in both data sources DB1 and DB2.

4 Methods and principles

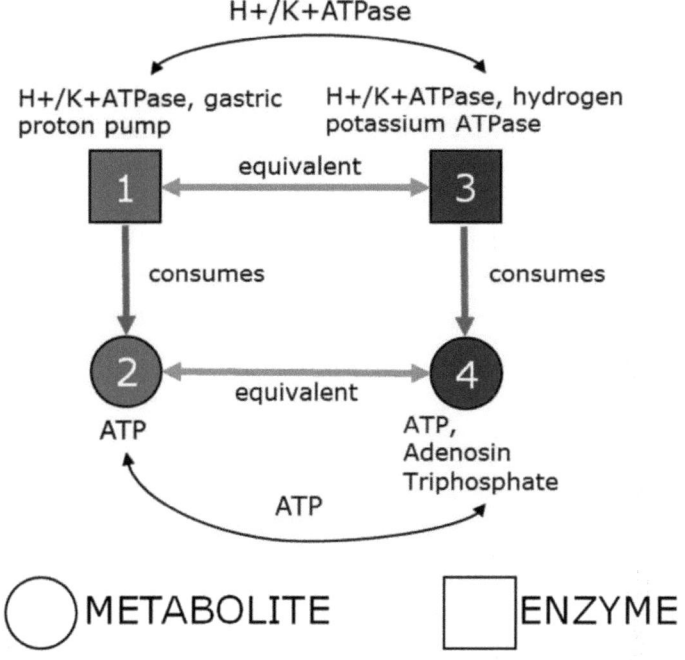

FIGURE 22 WORKED EXAMPLE FOR STRUCTALIGN. DIFFERENT SHADES ARE USED TO DISTINGUISH DATA SOURCES. NODE SHAPE REPRESENTS DIFFERENT CLASSES OF CONCEPTS, SQUARE FOR ENZYMES AND CIRCLE FOR METABOLITES. ROUND ARROWS SHOW MATCHING SYNONYMS, WHEREAS VERTICAL ARROWS REPRESENT EXISTING KNOWLEDGE FROM DATA SOURCES AND HORIZONTAL ARROWS ARE CREATED BY STRUCTALIGN.

StructAlign starts to consider the first pair of concepts, here concept 1 and 3, which share at least one preferred synonym (H+/K+ATPase) and are of the same concept class (enzyme). The reachability list of concept 1 includes concept 2 and the reachability list of concept 3 includes concept 4. The undirected breadth-first search of StructAlign will find concept 2 and 4 both being present at depth 1. As concept 2 and 4 share at least one preferred synonym (ATP) and are of the same concept class (metabolite), StructAlign collected enough evidence to create a new relation (green) between concepts 1 and 3. In the next step StructAlign proceeds to the next pair of concepts 2 and 4 between DB1 and DB2, which share at least one preferred synonym and will map them as being equivalent (green) because of the name match present between concepts 1 and 3.

4.2.2.3 OTHER DATA INTEGRATION METHODS

In addition to the mapping methods presented afore and evaluated in this study, the following selection of mapping methods shows how other information can be incorporated to deduce new relationships between concepts. This functionality is similar to that seen in "Biozon" (2.3.2). A more complete list of data integration methods can be found in 9.2 "Data integration methods in ONDEX".

Transitive mapping: Transitive relationships between concepts are inferred from existing relations. For example, if concept A is identified to be equivalent to concept B and concept B is known to be equivalent to concept C, then a new equivalent relationship between concept A and concept C is created by this mapping method.

Sequence2pfam mapping: The assignment of protein domain functional information to protein sequences is achieved by exporting the sequence data into a FASTA [77] file and matching against the consensus sequences from a local PFAM database [78] using BLAST [79] or HMMER [80]. The results are used to create relations between concepts representing proteins and relevant entries in representation of the PFAM database.

External2go mapping: The GO consortium provides reference lists of GO terms that map terms to other classification systems, for example EC [70] enzymes or PFAM domains. The *external2go* mapping parses these lists and creates relations between entries of the GO database and entries of the other classification system.

These few examples together with the methods listed in 9.2 illustrate the wide range of information which is utilized by mapping methods in ONDEX including simple name matches, sequence similarity search, orthology prediction, graph-pattern matching and even complex text mining based information retrieval. Furthermore it is not difficult to add new mapping methods to ONDEX.

4.2.2.4 EVALUATION METHODS

The mapping algorithms presented here can be configured using different parameters. According to the selection of the parameters these methods yield different mapping results. To evaluate their behaviour, two different test scenarios were used: the mapping of equivalent elements in ontologies and the integration and analysis of metabolic pathways.

The evaluation of a mapping method requires the identification of a reference data set, sometimes also referred to as a "gold standard", describing the links that really exist between data and that can be compared with those which are computed. Unfortunately, it is rare that any objective definition of a "gold standard" can be found when working on biological data sets and so inevitably most such evaluations require the development of expertly curated data sets. Since these are time-consuming to produce, they generally only cover relatively small data subsets and therefore the evaluation of precision and recall is inevitably somewhat limited.

In Section 4.2.3.1 "Mapping methods – Enzyme Nomenclature vs. Gene Ontology " the results of mapping together two ontologies; namely the Enzyme Commission (EC) Nomenclature [70] and Gene Ontology (GO) [71] are presented. In this case, the Gene Ontology project provides a manually curated mapping to the Enzyme Nomenclature called *ec2go*. Therefore *ec2go* has been selected as the first gold standard. The cross references between the two ontologies contained in the integrated data were also considered as the second gold standard for this scenario.

Section 4.2.3.2 "Mapping methods – KEGG vs. AraCyc" presents the results from the evaluation of a mapping created between the two metabolic pathway databases KEGG and AraCyc. Unfortunately, a manually curated reference set is not available for this scenario. Therefore it was necessary to rely on the cross references between the two databases that can be calculated through accession based mapping as the nearest equivalent of a gold standard for this scenario.

4.2.3 RESULTS

The mapping algorithms were evaluated using the standard measures of precision (Pr), recall (Re) and F_1-score [81]:

$$Pr = \frac{tp}{tp+fp} \quad Re = \frac{tp}{tp+fn} \quad F_1 = \frac{2*Pr*Re}{Pr+Re}$$

The accession based mapping algorithm (Acc) was used with default parameters, i.e. only using non-ambiguous accessions. This choice has been made to obtain a "gold-standard" through accession based mapping, i.e. increasing the confidence in the relations created. When evaluating the synonym mapping (Syn) and StructAlign (Struct) algorithms, parameters were varied to examine the effect of the number of synonyms that must match for a mapping to occur. This is indicated by the number in brackets after the algorithm abbreviation (for example Struct(1)). A second variant of each algorithm in which related synonyms of concepts were used to find a mapping was also evaluated. The use of this algorithmic variant is indicated by an asterisk suffix on the algorithm abbreviation (for example Syn(1)*).

4.2.3.1 MAPPING METHODS – ENZYME NOMENCLATURE VS. GENE ONTOLOGY

The goal of this evaluation was to maximise the projection of the Enzyme Commission (EC) nomenclature onto the Gene Ontology. This would assign every EC term one or more GO terms. For the evaluation ec2go (revision 1.67) and gene_ontology_edit.obo (revision 5.661) obtained from ftp://ftp.geneontology.org was used. Additionally enzclass.txt (last update 2007/06/19) and enzyme.dat (release of 2008/01/15) were downloaded from ftp://ftp.expasy.org . The data files were parsed into the ONDEX data structure and the mapping algorithms applied using the ONDEX integration pipeline. To determine the optimal parameters for this particular application case different combination of the mapping algorithms with the

4 Methods and principles

variants and parameter options as described above have been systematically tested. Table 6 summarises the mapping results and compares the performances with the "gold standards" data sets from ec2go and by accession mapping (Acc).

TABLE 6 MAPPING RESULTS FOR ENZYME NOMENCLATURE TO GENE ONTOLOGY

Method	TP,FP ec2go	TP,FP Acc	Pr,Re [%] ec2go	Pr,Re [%] Acc	F_1-score ec2go	F_1-score Acc
Ec2go	8063 , 0	8049 , 14	100 , 100	99 , 84	100	91
Acc	8049 , 1441	9490 , 0	84 , 99	100 , 100	91	100
Syn(1)	7460 , 934	7462 , 932	88 , 92	88 , 77	**90**	**82**
Syn(1)*	7605 , 2581	7606 , 2580	74 , 94	74 , 80	83	77
Syn(2)*	4734 , 374	4738 , 370	92 , 58	92 , 49	71	64
Syn(3)*	2815 , 117	2816 , 116	96 , 34	96 , 29	51	45
Struct(1)	1707 , 63	1712 , 58	96 , 21	96 , 18	34	30
Struct(1)*	1761 , 279	1766 , 274	86 , 21	86 , 18	34	30
Struct(2)	7460 , 934	7462 , 932	88 , 92	88 , 77	**90**	**82**
Struct(2)*	7605 , 2581	7606 , 2580	74 , 94	74 , 80	83	77
Struct(3)	7460 , 934	7462 , 932	88 , 92	88 , 77	**90**	**82**
Struct(3)*	7605 , 2581	7606 , 2580	74 , 94	74 , 80	83	77

ec2go = imported mapping list (1st gold standard), Acc = Accession based mapping (2nd gold standard), Syn = Synonym mapping, Struct = StructAlign, * = allow related synonyms, TP = True Positives, FP = False positives, Pr = Precision, Re = Recall, F_1-score. Synonym mapping was parameterised with a number that states how many of the names had to match to create a link between concepts. StructAlign was parameterised with the depth of the graph neighbourhood.

The first two rows of Table 6 show the performance of the "gold standard" methods tested against themselves. As can be seen by reviewing the F_1 scores in the subsequent rows of Table 6, the most accurate synonym mapping requires the use of just one synonym. It does not help to search for further related synonyms (Syn(1,2,3)*). The explanation for this is that the EC nomenclature does not distinguish between exact and related synonyms. Therefore, concepts belonging to the EC nomenclature have only one preferred concept name (exact synonym) arbitrarily chosen to be the first synonym listed in the original data sources. A large number of entries in the EC nomenclature only have one synonym described, which probably explains the low recall of Syn(2)* and Syn(3)*.

4 Methods and principles

The use of the more complex StructAlign algorithm, which uses the local graph topology to identify related concepts has low recall when only a single synonym is required to match and a depth cut-off of 1 is used (Struct(1) and Struct(1)*). This almost certainly results from differences in graph topology between EC nomenclature and Gene Ontology. The Gene Ontology has a more granular hierarchy, i.e. there is more than one hierarchy level between two GO terms mapped to EC terms, whereas the EC terms are only one hierarchy level apart. As the StructAlign depth cut-off search parameters are increased, more of the graph context is explored and accordingly the F_1 scores improved.

The highest F_1-scores come from Syn(1), Struct(2) and Struct(3) respectively. Allowing the related synonyms into the search (the * algorithm variants) did not improve precision. Neither did extending the graph neighbourhood search depth from Struct(2) to Struct(3) as all the neighbourhood matches had already been found within search depth 2.

During the integration of these datasets for this evaluation some inconsistencies in the ec2go mapping list have been observed. The identification of such data quality issues are often a useful side-effect of developing integrated datasets. The inconsistencies identified are listed in Table 7 and were revealed during the import of the ec2go data file after preloading the Gene Ontology and EC nomenclature into ONDEX.

Presumably most of the problems are due to the parallel development of both ontologies, i.e. GO references that were transferred or EC entries being deleted or vice versa. A few of the inconsistencies were possible typo errors. It remains a possibility that other "silent" inconsistencies are still in ec2go that these integration methods would not find.

TABLE 7 INCONSISTENCIES IN EC2GO

Accession	Mapping	Reason for failure
GO:0016654	1.6.4.-	Enzyme class does not exists, transferred entries
GO:0019110	1.18.99.-	Enzyme class does not exists, transferred entries
GO:0018514	1.3.1.61	Enzyme class does not exists, deleted entry
2.7.4.21	GO:0050517	GO term obsolete
GO:0047210	2.4.1.112	Enzyme class does not exists, deleted entry
1.1.1.146	GO:0033237	GO term obsolete
GO:0016777	2.7.5.-	Enzyme class does not exists, transferred entries
GO:0004712	2.7.112.1	Enzyme class does not exists, possible typo
2.7.1.151	GO:0050516	GO term obsolete

Every inconsistency was checked by hand against gene_ontology_edit.obo, enzclass.txt and enzyme.dat.

A more recent analysis of data files downloaded on 09/07/2009 revealed that the above presented errors still persist. The number of obsolete GO terms in ec2go changed from 3 to 134, this might indicate a systematic error or intension by the curators of the ec2go mapping list. Beside this significant change in numbers, the only two other inconsistencies identified were:

- GO:0047193 to 2.3.1.70 (Deleted entry.), new inconsistency
- GO:0004712 to 2.7.112.1 (spelling mistake in GO xref, should be EC:2.7.12.1), see above

4.2.3.2 MAPPING METHODS – KEGG VS. ARACYC

The KEGG and AraCyc pathway resources are both valuable for biologists interested in metabolic pathway analysis. Due to the different philosophies behind these two databases [82], however, they do have differences in their contents. Plant scientists wishing to work with biochemical pathway information would therefore benefit from an integrated view of AraCyc and the *A. thaliana* subset of KEGG and so this makes a realistic test. These two databases were chosen for this evaluation because both pathway databases annotate metabolites and enzymes in the pathways with standardised CAS registry numbers [83] and ATG numbers [84] respectively. It is therefore possible to evaluate the precision, recall and F_1-score of the different mapping methods using accession based mapping between these registry numbers as a "gold standard".

For this evaluation the KEGG database (release 46) obtained from ftp://ftp.genome.jp and the AraCyc database (release 4.1) obtained from http://www.arabidopsis.org/biocyc/index.jsp were used. The AraCyc database contained 1719 metabolites and 6192 enzymes. The KEGG database contained 1379 metabolites and 1126 enzymes. The evaluation results from the mapping between metabolites from these two databases is summarised in Table 8.

TABLE 8 MAPPING RESULTS FOR KEGG AND ARACYC DATABASES – METABOLITES

Method	TP	FP	Pr [%]	Re [%]	F_1-score
Acc	662	0	100	100	100
Syn(1)	516	726	41	77	54
Syn(1)*	600	1038	36	90	52
Syn(2)	14	0	100	2	4
Syn(2)*	396	348	53	59	**56**
Syn(3)*	190	114	62	28	39
Struct(2)	374	448	45	56	50
Struct(2)*	416	614	40	62	49
Struct(3)	382	470	44	57	**50**
Struct(3)*	432	670	39	65	48

Acc = Accession based mapping (gold standard), Syn = Synonym mapping, Struct = StructAlign, * = allow related synonyms, TP = True Positives, FP = False positives, Pr = Precision, Re = Recall, F1-score. Synonym mapping was parameterised with a number that states how many of the names had to match to create a link between concepts. StructAlign was parameterised with the depth of the graph neighbourhood

Accession based mapping between metabolite names found 662 out of 1379 possible mappings. A closer look reveals that CAS registry numbers are not always assigned to metabolite entries. Therefore, the accession based mapping misses possible links and cannot be used naively as a gold standard for this particular application case. In this evaluation, accession based mapping underestimates possible mappings, which leads to low precision for synonym mapping and StructAlign. A random set of the false positive mappings returned by Syn(2)* and Struct(3) has been manually reviewed and this revealed that most of the mappings were correct and metabolites shared similar chemical names. Subject to further investigation, this example shows that relying only on accession based data for integration might miss out some important links between data sources.

The evaluation results from the mapping between enzyme names from KEGG and AraCyc are summarised in Table 9.

TABLE 9 MAPPING RESULTS FOR KEGG AND ARACYC DATABASES – ENZYMES

Method	TP	FP	Pr [%]	Re [%]	F_1-score
Acc	2240	0	100	100	100
Syn(1)	282	0	100	12	22
Syn(1)*	2234	344	86	99	**92**
Syn(2)*	144	2	98	6	12
Syn(3)*	6	0	100	1	1
Struct(1)	234	0	100	10	18
Struct(1)*	2134	0	100	95	**97**
Struct(2)	282	0	100	12	22
Struct(2)*	2232	336	86	99	92
Struct(3)	282	0	100	12	22
Struct(3)*	2234	336	86	99	**92**

Acc = Accession based mapping (gold standard), Syn = Synonym mapping, Struct = StructAlign, * = allow related synonyms, TP = True Positives, FP = False positives, Pr = Precision, Re = Recall, F1-score. Synonym mapping was parameterised with a number that states how many of the names had to match to create a link between concepts. StructAlign was parameterised with the depth of the graph neighbourhood

The accession based mapping between enzymes uses the ATG numbers available in both KEGG and AraCyc. Entries from KEGG can be labelled with two or more ATG numbers representing multiple proteins involved in the enzymatic function, whereas AraCyc entries usually only have one ATG number. This results in one-to-many hits between KEGG and AraCyc explaining why a total of 2240 instead of only 1126 mapping were found. This is a good example of how the differences in the semantics between biological data sources make it difficult to define a gold standard for evaluating data integration methods.

The key finding from this evaluation based on mapping enzyme names is that the more flexible synonym and StructAlign algorithms, in particular those using the related synonyms variant, show very equivalent precision to the more simple accession mapping methods. Therefore, these methods can be considered as practical alternatives when direct accession mapping is not possible between two data sources. Furthermore, flexible mapping methods which explore a larger search space among concept names and synonyms have the potential to improve the recall when aligning metabolic networks without unduly sacrificing precision.

4.2.3.3 VISUALISING RESULTS

Data integration involving large data sets can create very large networks that are densely connected. To reduce the complexity of such networks for the user, information filtering, network analysis and visualisation (see Figure 21, step 3) is provided in a client application for ONDEX called OVTK [2]. The combination of data integration and graph analysis and visualisation has been shown to be valuable for a range of data integration projects in different domains, including microarray data analysis [2], support of scientific database curation [85, 86] and assessing the quality of terms and definitions in ontologies such as the Gene Ontology [87].

A particularly useful feature in the OVTK is to visualise an overview of the types of data that have been integrated in the ONDEX system. This overview is called the ONDEX meta-graph. It is generated as a network based on the integration data structure used in ONDEX, which contains a type system for concepts and relations. Concepts are characterised using a class hierarchy and relations have an associated type (see 4.1 "ONDEX integration data structure"). This information about concept classes and relation types is visualised as a graph with which the user can interact to specify semantic constraints – such as changing the visibility of concepts and relations in the visualisation and analysis of the integration data structure.

As an illustration, the integration of KEGG and AraCyc for this evaluation results in more than 100,000 concepts and relations. The mapping methods were run with optimal parameters identified in Section 4.2.3.2 "Mapping methods – KEGG vs. AraCyc". After filtering down to a specific pathway using methods available in the ONDEX visualisation interface OVTK, it was possible to extract information from the integrated data as presented in Figure 23.

Figure 23 displays parts of the *chlorophyll a biosynthesis I* pathway from AraCyc (upper part of Figure 23a) mapped to the corresponding subset from KEGG (lower part of Figure 23a). It is now possible to visualise the differences between the two integrated pathways. AraCyc provides more metabolites (octagon, for example *phytyl-PP*) on the reaction (star, *RXN1F-66*). KEGG denotes also the reverse direction of the reaction *chlorophyllide* to *chlorophyll a*.

The meta-graph is shown in Figure 23b. This visualisation shows that the integrated dataset consists of pathways (Path), reactions (Reaction) belonging to these pathways ("member is part"), metabolites (Comp) consumed ("cs_by") or produced ("pd_by") by the reactions and several combinations of enzymes (Enzyme) and proteins (Proteins) catalysing ("ca_by") the reactions. The meta-graph provides the user with a useful high level overview of the conceptual schema for this integrated data.

4 Methods and principles

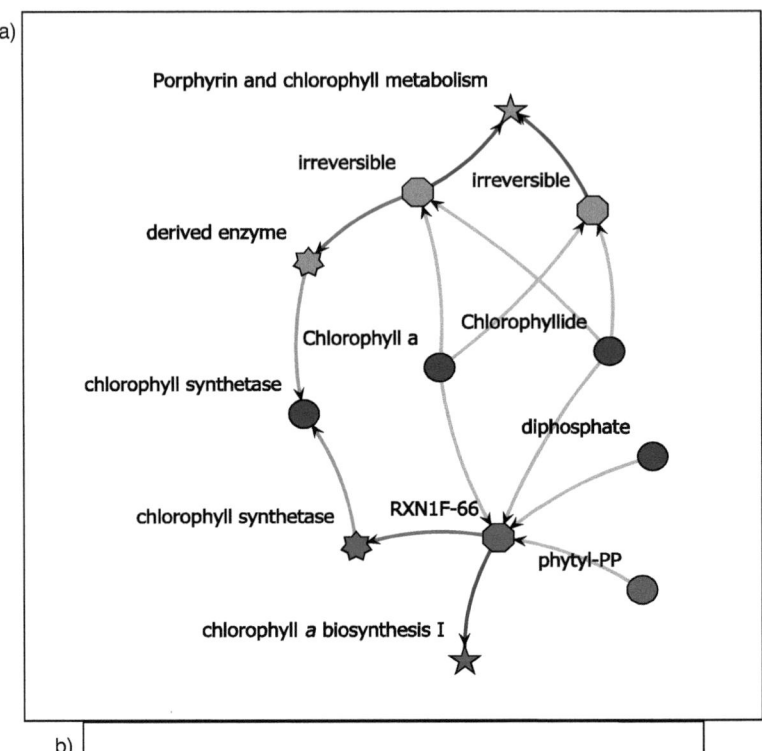

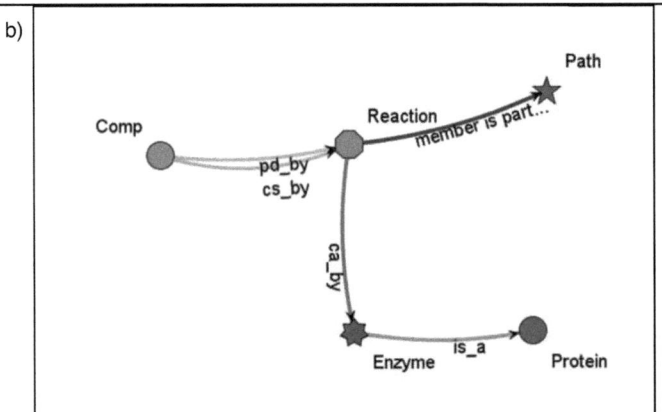

FIGURE 23 A) CHLOROPHYLL A BIOSYNTHESIS I PATHWAY FROM ARACYC (BOTTOM NODES) WITH CORRESPONDING SUBSET FROM KEGG (TOP NODES). NODES IDENTIFIED TO BE EQUIVALENT IN DARKER SHADE. B) META-GRAPH PROVIDING AN OVERVIEW OF THE INTEGRATED DATA, NODE COLOUR AND SHAPE DISTINGUISHES CLASSES; EDGE COLOUR DISTINGUISHES DIFFERENT RELATION TYPES.

4.2.4 Discussion

Alternative methods for creating cross-references (mappings) between information in different but related data sources have been presented. This is an essential component in the integration of data having different technical and semantic structures. Two realistic evaluation cases were used to quantify the performance of a range of different methods for mapping between the concept names and synonyms used in these databases. A quantitative evaluation of these methods shows that a graph based algorithm (StructAlign) and mapping through synonyms can perform as well as using accession codes. In the particular application case of linking chemical compound names between pathway databases, the StructAlign algorithm outperformed the most direct mapping through accession codes by identifying more elements that were indirectly linked. Manual inspection of the false positive mappings showed that both StructAlign and synonym mapping methods can be used where accession codes are not available to provide links between equivalent data source concepts. The combination of all three mapping methods yields the most complete projection between different data sources. This is an important result because it is not always possible to find suitable accession code systems that provide the direct cross-references between databases once you move outside the closely related data sources that deal with biological sequences and their functional annotations.

A similar approach to StructAlign called "SubTree Match" has been described in [88] for aligning ontologies. This work extends this idea into a more general approach for integrating biological networks and furthermore, presents a formal evaluation in terms of precision and recall.

A particular challenge in this evaluation has been to identify suitable "gold standard" data sets against which to assess the success of the algorithms developed. The results presented here are therefore not definitive, but represent the best quantitative comparison that could be achieved in the circumstances. Therefore these results represent a pragmatic evaluation of the relative performance of the different approaches to concept name matching for data integration of life science data sources.

4.3 EXCHANGING INTEGRATED DATA

Parts of this chapter have been published in the Journal of Integrative Bioinformatics in 2007, see [7].

4.3.1 MOTIVATION

The final step of data integration in ONDEX is to communicate results between parts of the system, scientific peers or other data sinks (see Figure 24). For this task it is necessary to define an appropriate transport or exchange format [7]. Although many different approaches to data integration exist (see Chapter 2), which have been reviewed earlier, there are no standard formats, specifically designed for the exchange of integrated datasets. Thus, users of data integration systems have to rely on the proprietary interfaces and exchange formats from the different data integration platforms. The use of XML, RDF and OWL provides a formal framework for the exchange of data. Many formats based on XML, RDF and OWL have been developed for the exchange of biological data. However, none of the existing XML, RDF and OWL based formats alone is suitable as a generic domain independent exchange format for integrated datasets (**fifth challenge**, see 3.2 "Challenges for data integration").

Traditionally, data sources have been distributed using proprietary flat-file formats or tab delimited database dumps. It is a popular myth that the appearance of XML has made these formats obsolete, and experience shows that still only about 5% of all data sources provide an XML based format [53]. Several data sources are still exclusively distributed in proprietary flat-file formats or as database dumps. A high percentage of these data sources provide no exchange format and can only be accessed through HTML based web-pages. Standardised exchange formats have great potential in improving and simplifying data integration, as generic tools and interfaces can be (re-)used. Widespread adoption of an exchange format leads to improved data documentation and will inevitably improve the exchange formats as formal agreements on data content and level of detail are reached between data providers.

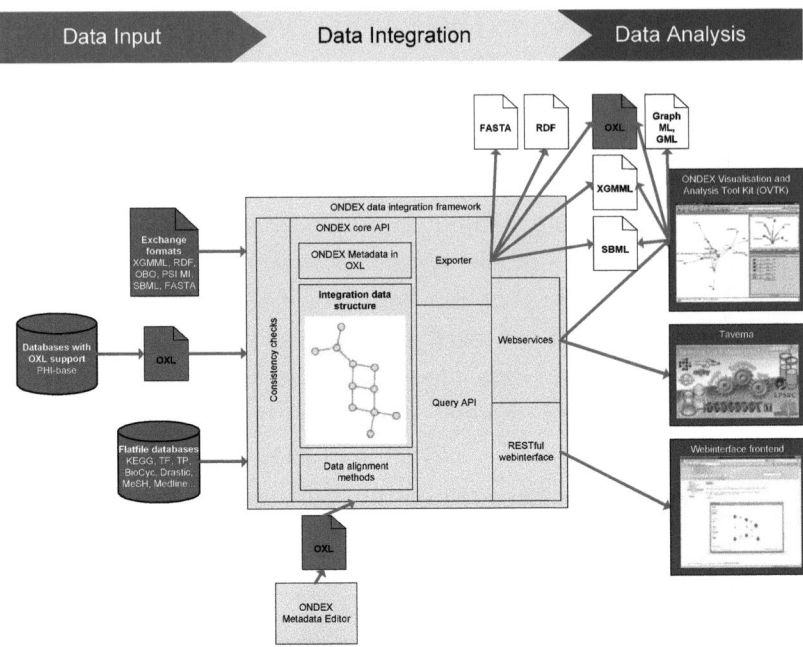

FIGURE 24 SHOWS THE DIFFERENT COMPONENTS OF ONDEX AND HOW ONDEX MAKES USE OF THE NEW EXCHANGE FORMAT (OXL HIGHLIGHTED) AS WELL AS OTHER STANDARD EXCHANGE FORMATS. INTEGRATION RUNS CONSISTS OF THREE STEPS: INPUT OF DATA FROM DIFFERENT DATA SOURCES (INCLUDING OXL), THE INTEGRATION PROCESS IN THE ONDEX INTEGRATION DATA STRUCTURE (INITIALISATION DATA VIA OXL), AND DATA ANALYSIS USING DIFFERENT TOOLS AND INTERFACES (DATA EXCHANGE USING OXL).

4.3.2 REQUIREMENTS FOR EXCHANGING INTEGRATED DATA SETS

Several XML-based exchange formats have been developed for representing data and models in specific biological areas, including the "Architecture for Genomic Annotation, Visualization and Exchange" (AGAVE) [89], "Biological Pathway Exchange" (BioPAX) format [90], "BIOpolymer Markup Language" (BIOML) [91], "Bioinformatics Sequence Markup Language" (BSML) [92], CellML Markup language for describing mathematical models [93], "Chemical Markup Language" (CML) [94], "Microarray Gene Expression Language" (MAGE-ML) [95], "Proteomics Experiment Markup Language" (PEML) [96], "Protein Markup Language"

(ProML) [97], "Proteomics Standards Initiative's Molecular Interaction" (PSI-MI) format [98], and "Systems Biology Markup Language" (SBML) [99].

TABLE 10 REQUIREMENTS FOR EXCHANGING INTEGRATED DATA WITH EXAMPLES

Motivation	Requirement	Example
Data integration needs to include all kinds of biological data, for example pathway data, gene expression data, biochemical reactions etc.	**i) Cover data from a broad range of application domains.**	Combined analysis of microarray and metabolomics data in the context of integrated pathway data.
Biological research progresses and new understanding may emerge that result in novel complex data structures.	**ii) Be extensible to combine many different complex data structures.**	The current transition from a gene-centric to network-based representation of molecular biology. [100]
Biological ontologies are subject to frequent changes.	**iii) Be flexible with meta data and semantical information from other sources.**	During the evolution of the PSI-MI format to the current release (version 2.5), major changes have occurred.
Not all relationships between biological entities are present in the data sources.	**iv) In addition to integrated data also include inferred information.**	New relationships between biological entities may be identified by data integration and analysis methods.
Integrated data may originate from several different data sources or be inferred computationally.	**v) Identify the original data source for integrated entities.**	Integrating several pathway databases, like KEGG and BioCyc at once.
Biological knowledge is steadily growing, as is the data contained in biological data sources.	**vi) Transport large amounts of integrated data.**	The KEGG database in its flat file representation is already more than 4GByte large.

Some of these exchange formats cover a broader biological domain than others. SBML, for example, is a language for describing models common to research in many areas of computational biology, including cell signalling pathways, metabolic pathways, gene regulation, and others [101]. SBML has become a *de facto* standard for representing formal quantitative and qualitative models at the level of biochemical reactions and regulatory networks [99]. BioPAX enables the integration of diverse pathway resources by defining an open file format specification for the exchange of biological pathway data. BioPAX includes representations for metabolic pathways, molecular interaction and promises future support for signalling pathways. It adopts some of the mechanism used in the PSI-MI format [90].

Several bioinformatics tools such as BCDE (BioPAX) [102], BioSpice (SBML) [103], Cellerator (SBML) [104], Cytoscape (BioPAX) [105], DBsolve (SBML) [106], E-Cell (SBML) [107], Gepasi (SBML) [108], Pathway Tools (BioPAX) [109], PATIKA (BioPAX) [110], StochSim (SBML) [111], Virtual Cell (SBML) [112], and VisANT (BioPAX) [113] use these formats for the importing/exporting of domain specific data. However, these formats are not usually used for exchanging a broad range of integrated data, which involves several data integration specific requirements. These requirements are listed in Table 10.

In biology, computational exchange languages are normally designed for a very specific application domain, such as the exchange of protein interactions (PSI-MI), description of microarray experiments (MAGE-ML), representation of formal quantitative and qualitative models (SBML) or exchanging pathway information (BioPAX); thus they do not satisfy the first requirement specified in Table 10.

Many of the above mentioned formats have only a limited functionality to incorporate new complex data structures (see second requirement in Table 10). Structural representation of complex data is predefined by the given format, for example the use of MathML [114] within the PSI-MI format to model equations. Adding complex data outside the defined representations therefore requires a change to the underlying schema of the format.

In defined ontology formats like BioPAX, which is represented in OWL-DL, all entity classes are predefined (may oppose third requirement in Table 10). For example if one likes to describe H+ as an inorganic substance (as subclass of *physicalEntity*) and not as an instance of the *smallMolecule* subclass of BioPAX requires the modification of the BioPAX schema to enable document validation by OWL reasoners.

Some formats, like MAGE-ML or PSI-MI provide functionality to include metadata about the information sources and methods used to generate contained information in a structured way. But not all of the above mentioned formats are able to satisfy the fourth and fifth requirement in Table 10 completely. The last requirement in Table 10 is also not always satisfied by the discussed exchange formats. One reason is that sophisticated document validation is involved. For some formats, especially OWL-DL based formats this is a demanding computational task with high inherent time and space complexity. From experience, this is the limiting factor for existing tools.

Therefore, none of the existing formats satisfy all of the aforementioned requirements for exchanging integrated biological data. A new exchange format had to be created to support the data integration, text mining and exchange of biological pathway data altogether as part of the ONDEX data integration framework.

4.3.3 THE OXL FORMAT

OXL was developed with the intension of exchanging integrated data sets between different components of the ONDEX system [2], and with external applications.

4.3.3.1 A BRIEF HISTORY OF OXL

The design of the OXL format is closely coupled with how data is represented in the integration data structure as defined in Section 4.1 "ONDEX integration data structure". The use of metadata facilitates covering data from a broad range of application domains (see first requirement in Table 10) and to be flexible with changes in metadata and semantic information from other sources (see third requirement in Table 10). Each concept and relation is marked with information on the data source (controlled vocabulary) from which it originates and the method that was used to create it in ONDEX (evidence) to keep track of provenance in the process of data integration (see fourth and fifth requirement in Table 10).

Prior to developing the OXL format, a careful investigation of existing formats such as SBML and BioPAX was performed. Using one of these formats for the ONDEX system would have had several advantages such as good tool support through, for example libSMBL (see http://www.sbml.org/software/libsbml/), the Jena API (see http://jena.sourceforge.net), and improved compatibility with other bioinformatics tools. Unfortunately, as reported in the previous section, despite SBML and BioPAX being well developed and successful exchange languages, they are not suitable as a generic exchange language to describe integrated data from multiple sources, or to exchange data between the different components of the ONDEX system. Given the obvious strengths of these data formats import and export functionality from ONDEX to a range of different formats, including certain aspects in SBML, is provided.

The overriding priority when selecting the OXL technical structure was a well-defined and widely adopted software-readable format (see last requirement in Table 10). XML, the eXtensible Mark-up Language [115] has been chosen because of its portability and increasingly widespread acceptance as the standard data language for bioinformatics [116]. There are, however, different approaches to data representation in XML including the RDF and OWL XML Schema.

Modelling the integration data structure in OWL would have restricted the outcome to predefined metadata for concepts and relations (here concept class and relation type): defined as OWL classes and sub classes. This would have imposed a static set of metadata at runtime. Although this is attractive in terms of the required programmatic complexity and computational reasoning; a different way, which would not require changing the OWL schema file every time new metadata for concepts or relations is introduced, seemed more reasonable. Also special tool support in the form of reasoners is required to make full use of the expressiveness of OWL-DL. Validation of documents in OWL-DL has inherent high time and space complexity. This tool support was missing and evaluation showed the same problems with OWL-DL as described by [117].

Tool support for the RDF format is better compared with OWL, because the complexity of document validation is reduced. The preferred way of representing relationships in RDF is in form of *subject – predicate – object*. Here *subject* and *object* correspond to ontology concepts and *predicate* to the relation between concepts. RDF does not allow direct association of complex metadata (complex relation types, evidence types etc.) with the *predicate*. One workaround is to model a relation as an *object*; references are added between the first participant in the relation to the *object* and the second participant in the relation. However, this reduces compatibility with existing tools, as this kind of modelling is not within the scope of standard RDF design. Representing ternary relationships in RDF is difficult and not an explicit part of the syntax elements. However, there exist several model variants for n-ary relations mentioned in (see http://www.w3.org/TR/swbp-n-aryRelations/). For cross system compatibility, an RDF based export for ONDEX has been developed by fellow student Keywan Hassani-Pak. Due to the aforementioned problems and other minor issues encountered, some information loss in the conversion to RDF is unavoidable.

4.3.3.2 OXL AS XML SCHEMA

The final conclusion was in favour of XML Schema because of its already widespread adoption and abundance of tools. XML Schema provides the most flexibility in modelling the integration data structure in XML. The principles used are reflected in the XML Schema of OXL as shown in Figure 25 to Figure 28. The start data element *ondex* includes either the *ondexmetadata* or the *ondexdataseq* element. This enables using one XML Schema for both, describing metadata (*ondexmetadata*) in terms of a controlled vocabulary, concept classes, relation types and evidences, and a complete graph structural representation for concepts and relations (*ondexdataseq*). An OXL file populated with metadata is required to initialise the ONDEX data integration framework with a common set of agreed-on metadata (see 5.1.2 "Formulating a consensus domain model in biology").

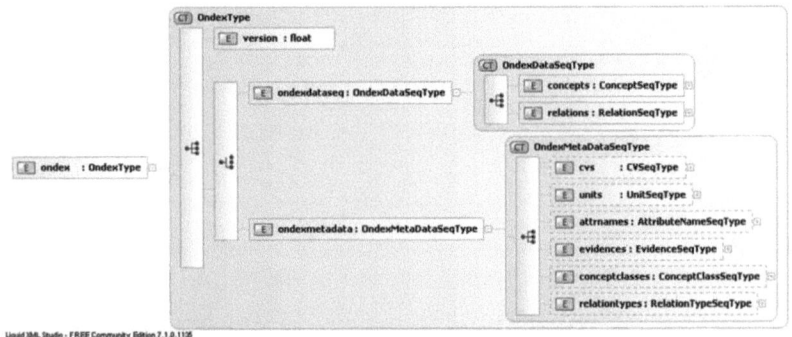

FIGURE 25 OXL XML SCHEMA – ONDEXDATA ELEMENT IS THE STARTING ELEMENT FOR OXL, THIS ELEMENT HAS DECISIVE CHARACTER ABOUT THE CONTENT OF THE OXL FILE, WHETHER DESCRIBING META-DATA OR AN ONDEX GRAPH.

The *ondexmetadata* element consists of an OndexMetaDataSeqType (see Figure 26), containing a list of all possible kinds of metadata used for the integration data structure (see third requirement in Table 10). Each metadata element is represented by a set of values containing: a unique identifier (*id*), human readable name (*fullname*) and free text description (*description*) to represent human readable information. These identifying values are common to all metadata elements, which are detailed in the following. The *cvs* element contains a list of data sources, called controlled vocabulary (*cv*). Types of *evidence* for concepts and relations are contained within the *evidences* element. The units of properties assigned to concepts and relations are grouped together in the *units* element. A *unit* can also be part of an attribute name (*attrname*) listed in the *attrnames* element. An attribute name is the first participant in a name-value pair, which together is termed a generalised data structure (GDS) [5] element; this can be assigned to any concept or relation. An *attrname* element has beside the common identifier set also a *datatype* element and a *specialisationOf* element. The *datatype* element usually specifies the JAVA (see http://java.sun.com) class of the GDS value, but is not limited to JAVA classes in general. This allows for representing complex data structures within a GDS, such as protein structure, or Position Weight Matrices (PWM), which describes the DNA binding motifs of transcription factors (see second requirement in Table 10). The *datatype* element intentionally avoids the use of predefined data types in XML Schema and thus enables addition of new data types without having to modify the XML Schema of OXL. The *specialisationOf* element can hold another *attrname* element, and thus represents the model taxonomy of attribute names. The same principle for modelling hierarchy within the metadata in OXL is implemented in both concept classes (*cc*) which are wrapped in the *conceptclasses* element, and relation types (*relation_type*) which are encompassed within the *relationtypes* element. A *relation_type* element has additional attributes to characterise the properties of the relation according to the OBO Relation Ontology [43]. A relation can have exactly one relation type. An example of the *pr_by* (preceded by) relations type is given in Figure 29, *pr_by* is a specialisation of the common root *r* (related) within the relation type hierarchy.

The *ondexdataseq* element is an OndexDataSeqType covering lists of concepts (*concepts*) and relations (*relations*). Each *concept* (see Figure 27) is identified by a unique *id* element (an integer). Textual information about a concept is held in *annotation* and *description*. The usual contract is that *annotation* is a short human readable definition for the concept and *description* can be a longer free text item associated with the concept. Additionally *concept* contains a *pid* element. This can be a short alternative textual identifier for the concept which is more understandable and usually assigned at creation time (parsing step). The *elementOf* element defines the controlled vocabulary (*cv*) or data source from which the concept originates. The *ofType* element represents the concept class (*cc*) for the concept.

4 Methods and principles

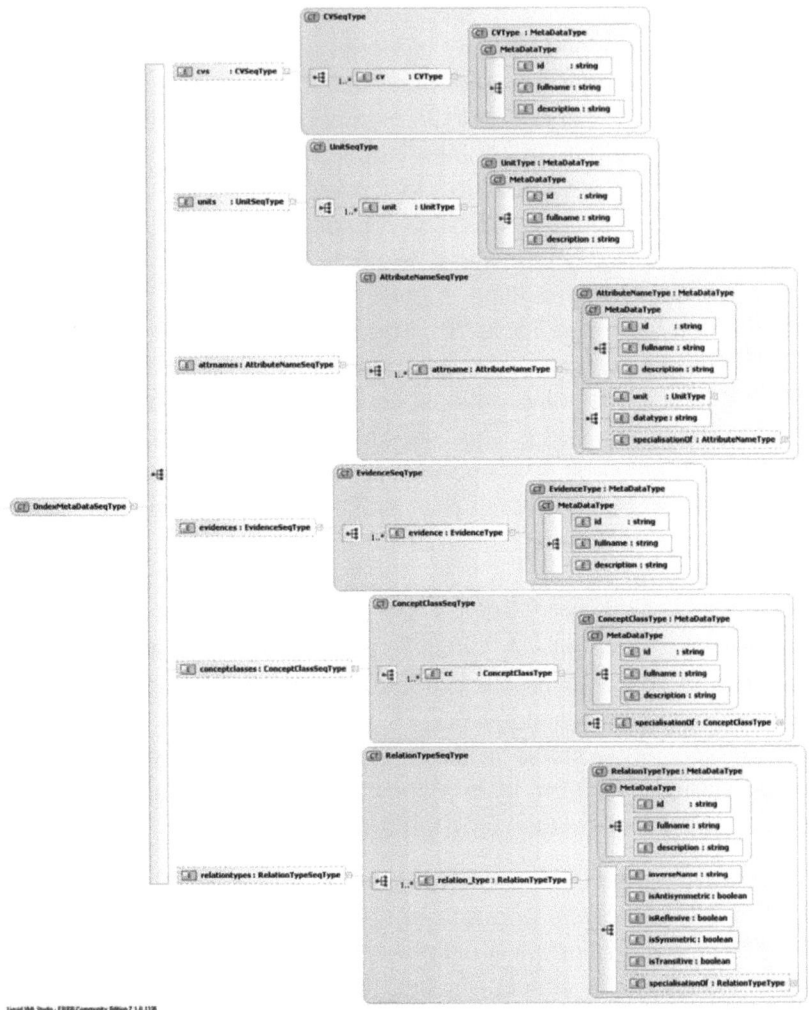

FIGURE 26 OXL XML SCHEMA – METADATA ELEMENT CONTAINS ALL CLASSES FOR META-DATA CURRENTLY PRESENT IN ONDEX, NAMELY CV, UNIT, ATTRIBUTENAME, EVIDENCETYPE, CONCEPTCLASS AND RELATIONTYPE.

The types of evidences (*evidence*) for the concept are collected within the *evidences* element. A concept has at least one type of evidence assigned. A concept can have synonyms which are expressed as concept names (*conames*), references to other data sources termed concept accessions (*coaccessions*), and arbitrary name-value pairs, encompassed by the generalised data structure (*cogds*). A *relation* (see Figure 28) is identified by the unique combination of *fromConcept*, *toConcept*, an optional *qualifier* concept, and a relation type (*ofType*). The additional properties *qualifier* enables OXL to represent ternary relationships. A relation is assigned evidence types (*evidences*) and arbitrary name-value pairs (*relgds*), which are handled in an equivalent way to concepts. Both concepts and relations contain a *contexts* element. The *contexts* element is a list of unique concept identifiers which constitute the context for that particular concept or relation. The list can be empty.

Except for the reuse of unique concept identifiers for *fromConcept*, *toConcept* and *qualifier* elements in a relation and *contexts* on concepts and relations, no other cross document references are made: all elements are always fully expanded, in a similar way to that used by the expanded form of the PSI-MI format [98]. Thus these elements always include a full copy of all sub-elements, even if such sub-elements like concept class or controlled vocabulary have been instantiated before. This facilitates the merging of several OXL documents and enables easy transformation of the OXL format into streamlined formats like HTML using XSLT style sheets. However, this introduces redundancy in the file format and thus increases the average size of OXL documents. Although even with the higher redundancy when compressed, these files are only marginally bigger than the equivalent non-redundant file format.

A versioning system of OXL is present to keep track of the changes within the document versions and provide scripts for upgrading existing data stored in OXL to the newest version. The format described here is the latest official release of the OXL format at the time of writing.

4 Methods and principles

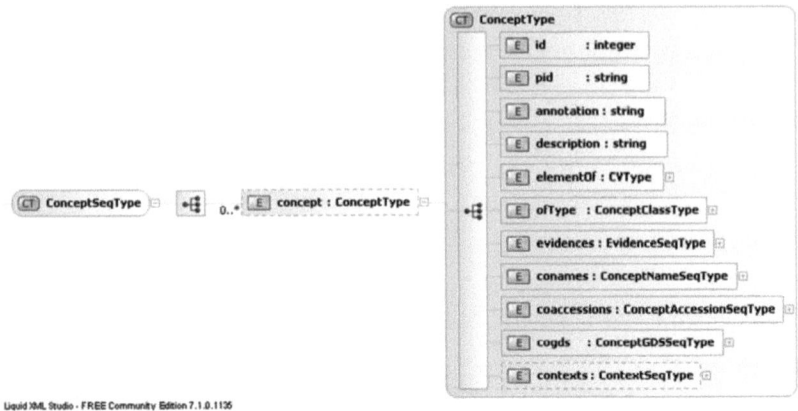

FIGURE 27 OXL XML SCHEMA – CONCEPT ELEMENT DESCRIBING PROPERTIES AND META-DATA OF ONE CONCEPT

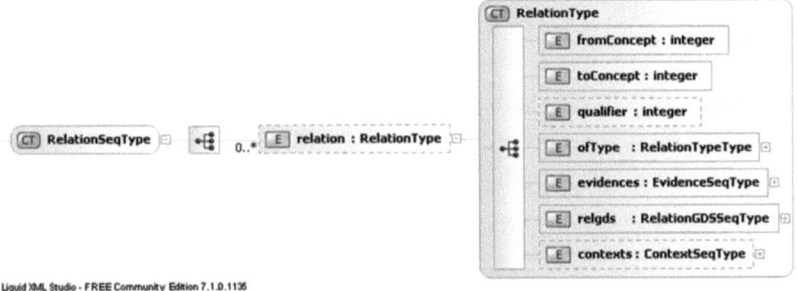

FIGURE 28 OXL XML SCHEMA – RELATION ELEMENT DESCRIBING PROPERTIES AND META-DATA OF ONE RELATION

```
<relation_type>.
    <id>pr_by</id>.
    <fullname>preceded by</fullname>.
    <description>A is the direct outcome of B</description>.
    <inverseName>none</inverseName>.
    <isAntisymmetric>false</isAntisymmetric>.
    <isReflexive>false</isReflexive>.
    <isSymmetric>false</isSymmetric>.
    <isTransitive>true</isTransitive>.
    <specialisationOf>.
        <id>r</id>.
        <fullname>r</fullname>.
        <description>is related to</description>.
        <inverseName>relation type</inverseName>.
        <isAntisymmetric>false</isAntisymmetric>.
        <isReflexive>false</isReflexive>.
        <isSymmetric>false</isSymmetric>.
        <isTransitive>false</isTransitive>.
    </specialisationOf>.
</relation_type>.
```

FIGURE 29 OXL METADATA EXCERPT FOR PR_BY RELATION TYPE, WHICH IS A SPECIALISATION OF THE MOST GENERAL RELATION TYPE "R"

4.3.3.3 SUPPORT FOR DATA INTEGRATION AND TEXT MINING

The OXL format was designed to support the tasks of data integration and text mining. Metadata has an important role in the course of integrating data in a semantically consistent way. Equivalent entities in different data sources, which represent the same biological object, for example genes imported from KEGG [65] and genes imported from BioCyc [66], share a common semantic definition, will also share a common metadata association (see second requirement in Table 10). Therefore, all data in OXL uses the same set of metadata compiled for the corresponding application domain (see third requirement in Table 10).

Provenance of data is tracked using controlled vocabulary (CV) and evidence types (see fourth and fifth requirement in Table 10). The controlled vocabulary marks the data source the data was imported from like KEGG [65], Transfac [118] or Transpath [119]. Instead of technically merging equivalent concepts and relations from different data sources, these concepts and relations are only aligned to each other (see 4.2 "Data alignment") using the special relation type equivalent (equ). This makes it possible to disentangle integrated data sets (see fifth requirement in Table 10). Evidence codes are used to keep track of how data was integrated. This facilitates the extraction or filtering of relations generated by a specified data analysis / integration method. Evidence codes can also be applied to inferred (generated) concepts and relations (see fourth requirement in Table 10).

4.3.3.4 TOOL SUPPORT AND APPLICATIONS OF OXL

OXL was developed as part of the ONDEX data integration framework. Thus it has a close relationship to the core integration data structure. There are three possible technical approaches for generating an OXL document. Firstly, a combination of a document objects model (DOM) and an XML writer/reader to handle OXL. This approach is appropriate for the conversion of small custom data models into OXL. Secondly, stream based XML parsing / writing techniques such as SAX (see http://www.saxproject.org/) or StAX (see http://www.jcp.org/en/jsr/detail?id=173) which can be used to directly read or write XML documents in the syntax of OXL. These techniques use the smallest memory footprint and are the only viable approach for large data sets. ONDEX uses StAX based parsing. Thirdly, for very large data models, the use of the ONDEX core API (*"core"*) is recommended. This API includes methods for the efficient construction and manipulation of very large graphs and OXL exports and imports. As ONDEX can store all data on disk in its own persistency layer and keeps only a cached subset of data in memory the handling of very large data sets is possible. Furthermore, the *"core"* can be used as a persistency management system for custom data integration applications in JAVA. The current ONDEX system includes support tools for reading and writing of data in the OXL format.

Editing metadata within the *"core"* is enabled through the Metadata Editor, which provides a graphical hierarchical-tree based representation of the metadata. For each application domain users or developers should be able to submit their own metadata additions and modifications to the central OXL document located in the Subversion repository of ONDEX, or to suggest such changes through the developer mailing list.

To demonstrate that OXL can be used for exchanging complete data sources without information loss, an OXL export for the Pathogen Host Interaction database (PHI-base) [86] was implemented by fellow student Rainer Winnenburg. This functionality is part of the current PHI-base release. It is possible to download the complete data source or selected query results in the OXL format; for example a search for the entry "PHI:441" yields the result table depictured in Figure 30. Figure 31 shows an excerpt from the generated OXL file and Figure 32 shows a screenshot of the ONDEX Visualisation Tool Kit (OVTK), with the loaded OXL file showing an organic layout of the data.

4 Methods and principles

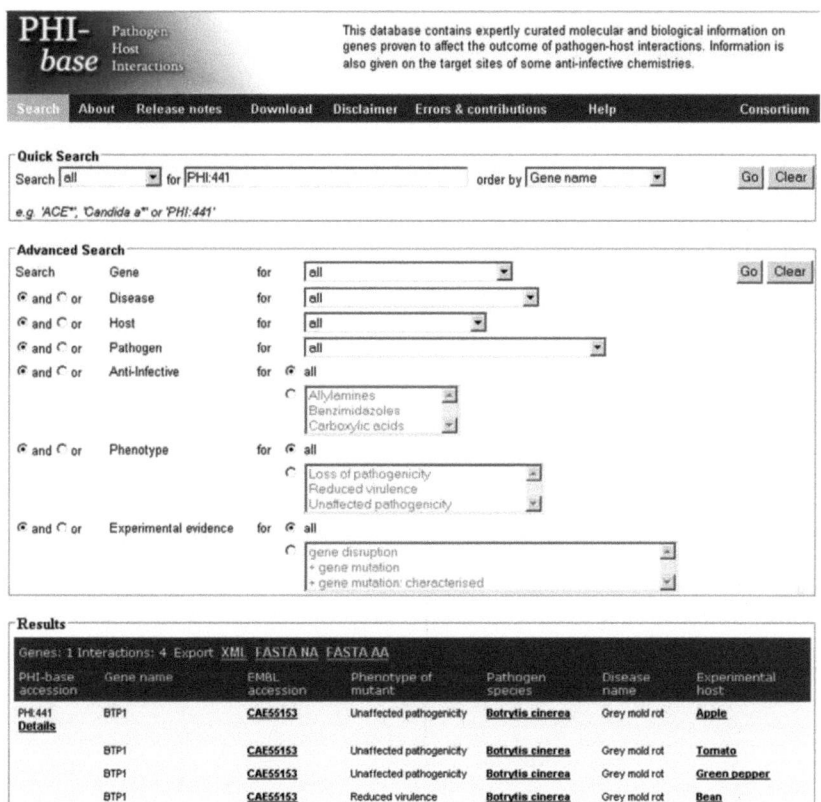

FIGURE 30 RESULT TABLE FOR PHI-BASE ENTRY PHI:441. TOP PART IS A QUERY INPUT FORM TO COMPOSE ADVANCED QUERIES ON PHI-BASE, WHEREAS THE BOTTOM PART DOES CONTAIN THE LIST OF RESULTS.

4 Methods and principles

```xml
<ondexdataseq>
  <concepts>
    <concept>
      <id>1</id>
      <pid>441</pid>
      <annotation />
      <description />
      <elementOf>
        <id>PHI</id>
        <fullname>PHI-base</fullname>
        <description>Pathogen - Host Interaction database</description>
      </elementOf>
      <ofType>
        <id>Protein</id>
        <fullname>Protein</fullname>
        <description>Protein</description>
      </ofType>
      <evidences>
        <evidence>
          <id>IMPD</id>
          <fullname>IMPD</fullname>
          <description>imported from database</description>
        </evidence>
      </evidences>
      <conames>
        <concept_name>
          <name>BTP1</name>
          <isPreferred>true</isPreferred>
        </concept_name>
      </conames>
      <coaccessions>
        <concept_accession>
          <accession>PHI:441</accession>
          <elementOf>
            <id>PHI</id>
            <fullname>PHI-base</fullname>
            <description>Pathogen - Host Interaction database</description>
          </elementOf>
          <ambiguous>false</ambiguous>
        </concept_accession>
```

FIGURE 31 EXCERPT OF OXL EXPORTED FROM PHI-BASE, HERE ONLY SHOWING THE FIRST CONCEPT WITH PHI-BASE ID PHI:441.

4 Methods and principles

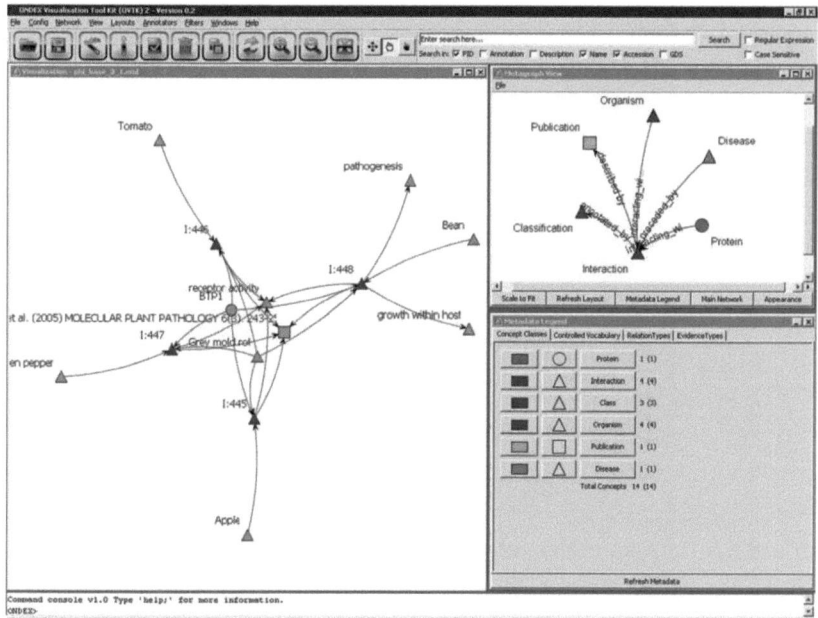

FIGURE 32 SCREENSHOT OF OVTK, DATA LOADED FROM OXL SHOWING RELATIONSHIPS OF PHI:441. ACTUAL GRAPH OF RELATIONSHIPS DISPLAYED IN CENTRAL "VISUALIZATION" FRAME. TOP RIGHT "METAGRAPH VIEW": OVERVIEW OF METADATA (CONCEPT CLASSES LIKE GENE, DISEASE, AND RELATION TYPES BETWEEN MEMBERS OF THESE CONCEPT CLASSES LIKE PRECEDED_BY, INTERACTING_WITH). LOWER RIGHT: CONCEPT CLASS COLOURS AND SYMBOLS.

4.3.3.5 THE ROLE OF OXL IN THE ONDEX DATA INTEGRATION FRAMEWORK

OXL can readily be used during the first step of the data integration process to load data from different data sources (see Figure 24) into the ONDEX core API (*"core"*), passing several consistency checks. The *"core"* uses ONDEX Metadata initialised with the help of an OXL document. Data alignment methods perform the data integration step from the original data into a semantical consistent view in the graph-based integration data structure (see 4.2 "Data alignment"). The data can be retrieved using the Query API and associated web services or by the prototype of a JSP based web interface. The web interface makes use of style sheets to transform OXL into HTML. The *"core"* also provides several exporters which work directly with the integration data structure. Once the data has been retrieved, it can in the third step be applied to data analysis. The ONDEX Visualisation Tool Kit (OVTK) is one such data analysis tool that can be applied to data in OXL format retrieved from the ONDEX system.

OXL is also used for data transfer between different components of the system and the storage of integrated datasets. Besides the OXL format, ONDEX makes use of a range of different exchange formats and data sources (see 9.1 "List of data formats supported by ONDEX"). Metadata for the data integration core is provided using OXL and can be easily edited using the Metadata Editor (see "Figure 35"). New relationships between imported concepts can be identified by data alignment methods like concept accessions matching, biological sequence similarity or text mining. After the data integration run has finished, the integrated data can be accessed through web services or exported directly into several exchange formats, including OXL, XGMML and SBML. The web services are used by the ONDEX Visualisation Tool Kit (OVTK) and Taverna [34]. A web interface prototype exists. Other applications can utilise one of the exchange formats provided by ONDEX to load data, for example XGMML can be used to load exported data from ONDEX into Cytoscape [105].

4.3.4 DISCUSSION

The decision to create a separate exchange format for integrated data sets is based on two main aspects. The first is that none of the existing bio-specific exchange languages were fully capable of satisfying all the requirements mentioned in Table 10 and that generic exchange formats like RDF and OWL without customisations would have imposed overly rigid restrictions on the integration data structure. Secondly, OXL was created to be closely linked with the integration data structure, so that if one would understand the integration data structure, then one will intuitively understand OXL.

4 Methods and principles

In addition to the native format OXL, the integration data structure can also be exported in RDF, in order to utilize existing RDF tools. This model involves workarounds detailed earlier. However, this format is less intuitive and requires more time and effort for others to use. Existing RDF tools such as Jena normally use an in-memory model, and as such users of the RDF format who need to process large amounts of data cannot benefit from the existing RDF tool support.

OXL satisfies all requirements listed in Table 10. The use of flexible metadata assignment to concepts and relations enables OXL to cope with a broad range of application domains (first requirement). Arbitrary complex data structures (second requirement) can be included using special name-value pairs, called generalised data structure (GDS). Metadata and references to information from other sources (third requirement) can easily be modified without changing the schema definitions. Inferred information (fourth requirement) and tracking of provenance (fifth requirement) is realised through special relation types and evidence types for concepts and relations. By using a StAX parsing approach, XML Schema validation and optional file compression it is possible to transport very large datasets in OXL (last requirement).

As shown in the example of PHI-base, exporting data sources into the OXL format is less work than developing a flat-file parser for the ONDEX system to import the data. Because OXL is almost fully expanded, with the exception of references to unique concepts IDs as used in a relation and for context, it is possible to create XSLT style sheets to transform OXL into other streamlined formats such as HTML. This principle is utilized by the ONDEX web interface prototype. Applications that want to make the most of integrated datasets created by ONDEX should use the OXL format. Other exchange formats like RDF, SBML or XGMML are also provided by the ONDEX system, but have inherent limitations as to how data can be represented and therefore may not contain all the information that would otherwise be contained within OXL.

The generation and exchange of integrated data from several sources also involves a legal aspect. Licensing models for data may differ between imported data sources, which makes it important to track provenance: from where the integrated datasets originate [53]. OXL includes such a provenance tracking mechanism. Concepts and relations from different data sources are only aligned to each other: concepts from two data sources that are equivalent remain as separate concepts within ONDEX and OXL, connected by an equivalence relation. It is therefore possible to limit the scope of information that is exchanged by disentangling integrated datasets and thus satisfy license agreements.

4.3.5 OUTLOOK

The ONDEX core API (*"core"*) has been released as a standalone JAVA module that can be reused by custom applications. This API includes support for reading and writing OXL files. The *"core"* also provides fast and efficient indexing and search functionality for OXL data. A versioning system for the OXL format and standardized curation for OXL metadata has been introduced using the SourceForge.net Subversion repository. This might encourage other data source providers to support the format. An extension to the ONDEX core would be a configurable importer for OXL, which downloads directly from SOAP based web services run by data sources supporting OXL.

5 DESIGN AND IMPLEMENTATION

This chapter highlights design decisions made for the implementation of the ONDEX system in an attempt to create a robust, usable and maintainable framework for data integration (**sixth challenge**, see 3.2 "Challenges for data integration"). First an overview of the design of the ONDEX system is given, highlighting certain aspects which address the mentioned challenges. The second part of this chapter presents implementation details of the integration data structure. In the last part of this chapter efficient techniques for querying the integration data structure and information retrieval are discussed.

5.1 SYSTEM DESIGN

The methods and principles of this thesis are implemented in the ONDEX data integration framework. A typical data integration pipeline for the ONDEX system is summarised in Figure 33. The ONDEX data integration framework consists of three main parts: 1.) Parsing heterogeneous data into the ONDEX integration data structure represented as an object based data model (see 4.1 "ONDEX integration data structure"); 2.) Identifying equivalent relations between entries of different data sources using integration methods (see 4.2 "Data alignment"); and 3.) Analysing integrated data using client tools (for example the ONDEX Visualisation Tool-Kit (OVTK)).

The ONDEX data integration framework is implemented in JAVA as platform independent software. In the current version of the ONDEX data integration framework, the integration data structure is implemented using an object based data model which makes use of the Berkley DB Java edition [120] for persistent data storage and the Lucene indexing engine (http://lucene.apache.org) for fast full text searches. With this configuration it is possible to import, index and search combinations of data sources that could result in millions of concepts.

5 Design and implementation

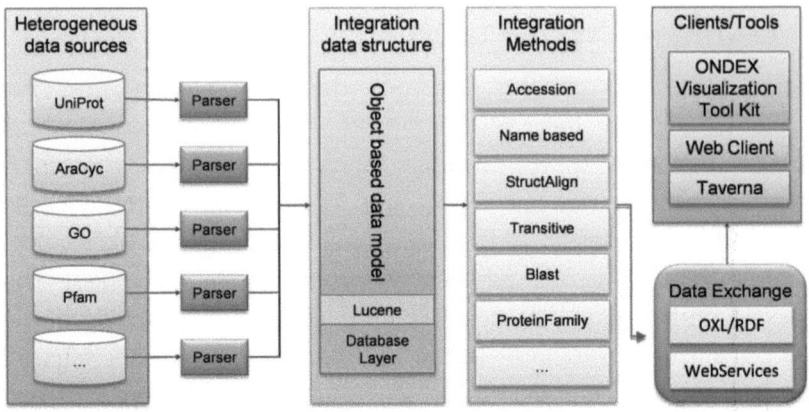

FIGURE 33 OVERVIEW OF A TYPICAL ONDEX PIPELINE CONSISTING OF THREE PARTS: 1) PARSING HETEROGENEOUS DATA SOURCES INTO THE OBJECT BASED DATA MODEL (INTEGRATION DATA STRUCTURE) OF ONDEX; 2) IDENTIFYING EQUIVALENT AND RELATED ENTRIES AND CREATING NEW RELATIONS BETWEEN THEM USING MAPPING METHODS; 3) ANALYSING THE INTEGRATED DATA USING CLIENT TOOLS, FOR EXAMPLE THE ONDEX VISUALISATION TOOL KIT (OVTK).

As the ONDEX system has been designed to be extensible, a new data source can easily be added by providing a new parser for it. For the complete list of current parsers see 9.1 "List of data formats supported by ONDEX". More sophisticated or faster integration methods, for example based on new sequence alignment tools, can also be added to the framework. For the complete list of current integration methods see 9.2 "Data integration methods in ONDEX". The web-services and the support for generic exchange formats (see 4.3 "Exchanging integrated data") allow for a multitude of analysis tools to be able to work with the resulting integrated data sets.

The implementation of the integration data structure as Generalized Object Data Model (see 5.2 "Implementing integration data structure") in the ONDEX framework already provides a large number of built-in consistency checks and integrated reporting facilities (based on Log4j, see http://logging.apache.org/log4j/). This supports the user during developing new features or just using the system by making the data integration process more verbose. Furthermore the core API of the ONDEX framework is exposed via web-services allowing the direct incorporation and linking with other web-service based frameworks like the text-mining suite U-Compare (see http://u-compare.org/) using the Taverna [34] workflow workbench.

5.1.1 KNOWLEDGE MODELLING AND DOMAIN INDEPENDENCE

Although the integration data structure has been primarily motivated by data integration in the life sciences, it is general enough to be applied to other domains of knowledge (**fifth challenge**) like social networks, patents mining, geo spatial data etc. whenever data can be represented as networks of concepts and relations. The integration data structure can be seen as a Meta Model which is instantiated using the appropriate Domain Knowledge (see Figure 34). Domain Knowledge has to be provided by a Domain Expert. After instantiation the integration data structure is able to act as a Domain Model (see 8.2.6 "Domain modelling").

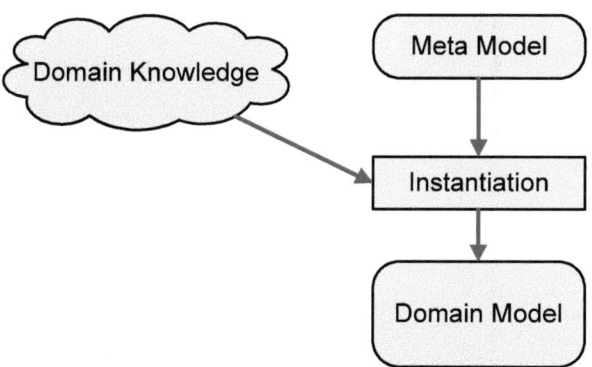

FIGURE 34 INSTANTIATION OF META MODEL WITH DOMAIN KNOWLEDGE LEADS TO DOMAIN MODEL

Each domain of knowledge requires its own set of semantics for the nodes and edges on the graph (see 4.1 "ONDEX integration data structure"). The instantiation of the integration data structure is achieved via an initialisation procedure to load metadata definitions from a dedicated configuration file in XML syntax (see 4.3

"Exchanging integrated data"). Additionally the JAVA API of the integration data structure supports the definition of metadata at runtime to enable changes or extensions to the domain definitions.

A graphical user interface based metadata editor (see Figure 35) supports the domain expert in the process of modelling the domain knowledge as metadata for the integration data structure. The metadata editor enables the domain expert to define hierarchies of Concept classes for semantics on concepts and hierarchies of Relation types for semantics on relations. Additionally lists of Attribute names, CVs, Evidence types and Units can be specified to describe additional attributes of concepts and relations. The resulting set of domain knowledge is saved in XML syntax.

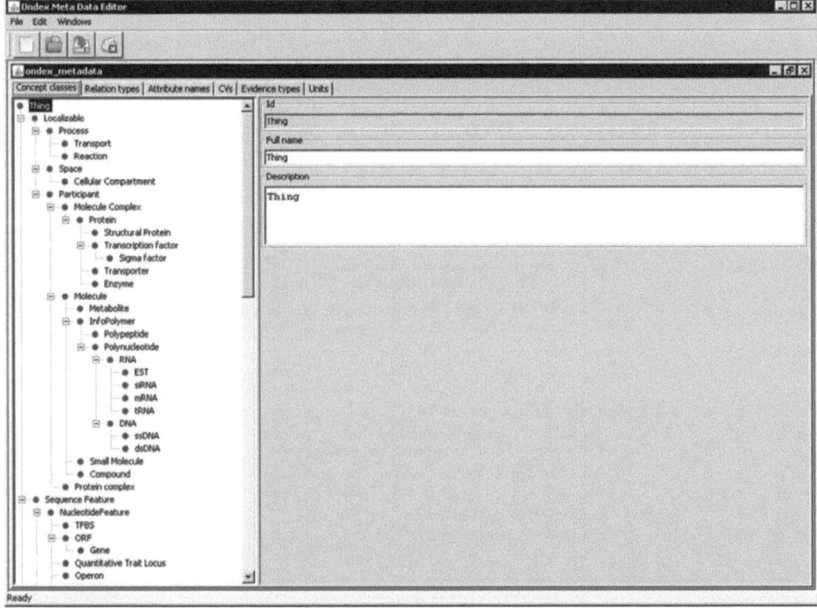

FIGURE 35 ONDEX META DATA EDITOR IMPLEMENTED BY JOCHEN WEILE. DIFFERENT ONDEX METADATA IS AVAILABLE ON SEPARATE TABS, WHEREAS THE HIERARCHY OF META-DATA IS DISPLAYED AS A TREE ON THE LEFT.

5.1.2 FORMULATING A CONSENSUS DOMAIN MODEL IN BIOLOGY

To make domain knowledge explicit is a difficult task. Even domain experts sometimes disagree when asked to give a detailed explanation of their domain of knowledge [121]. Several approaches to finding a top-level domain model for biology exists, for example BioTop [122] or GENIA [123]. Adapting such an approach has the advantage of providing a domain model with great detail covering many aspects of biology. Unfortunately at the same time this approach comes at a high price in terms of the complexity of the model.

The ONDEX approach to creating a domain model is more pragmatic and is based on reaching an agreement between users of a particular knowledge base and its use cases. To date most ONDEX applications have involved pathway information spanning major biological data source (for example KEGG, AraCyc, UniProt).

An important part of building a domain model is communicating it to intended users and visualisation of a domain model greatly assists model building and verification. In ONDEX a meta-graph is a visualisation of the domain model currently used in the integration data structure. It shows concepts classes as nodes and the respective relation types between them as edges. Different colours and shapes distinguish different kinds of concept classes and relation types. A meta-graph shares similarities with entity relationship (ER) diagrams used in computer science and provides a familiar visualisation to intended users.

Figure 36, Figure 37 and Figure 38 show the meta-graph of the databases AraCyc [73], KEGG [65] and UniProt [60]. All three meta-graphs differ because they represent different data sources and therefore different originating schemas. One prevailing commonality between the three meta-graphs is that proteomic data is always represented as concept class protein. As UniProt solely provides proteomic knowledge its overlap to the two other data sources is limited to this particular aspect. The agreement within the domain

5 Design and implementation

models between more similar data sources like AraCyc and KEGG is greater.

For pathway resources AraCyc and KEGG combined with UniProt database the following consensus domain model is presented in Figure 39:
- A Gene encodes one or more Proteins
- A Protein acts as an Enzyme
- An Enzyme has a Enzyme Classification
- An Enzyme catalyses one or more Reaction
- A Reaction is part of one or more Pathway
- A Reaction consumes or produces some Compound

Additional concept classes and relation types may as well exist depending on how the originating data model has been transformed into a domain model for the integration data structure. The greater the overlap or agreement between different knowledge domain models the easier and more complete data integration can be performed (see 4.2 "Data alignment").

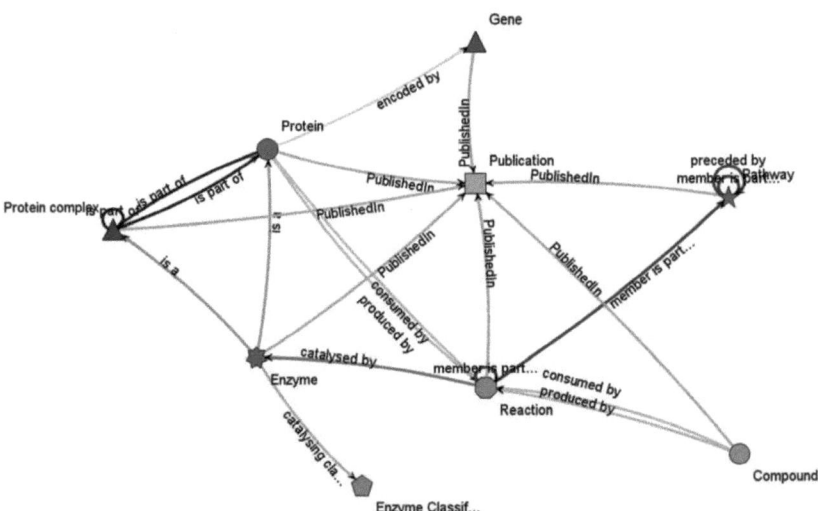

FIGURE 36 META-GRAPH ARACYC DATABASE CONTAINING CONCEPT CLASSES FOR PROTEINS (CIRCLE), REACTIONS (OCTAAGON) OR PATHWAYS (STAR) CONNECTED BY DIFFERENT RELATION TYPES DISTINGUISHED BY SHADES.

5 Design and implementation

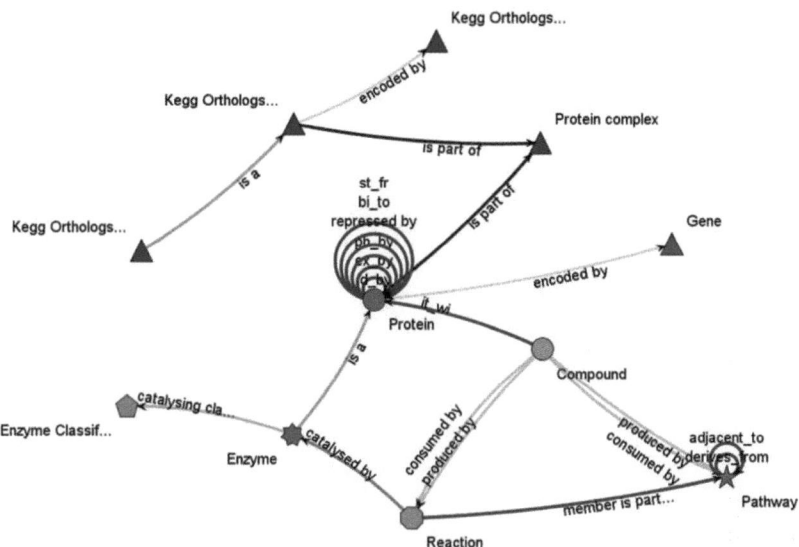

FIGURE 37 META-GRAPH KEGG DATABASE CONTAINING CONCEPT CLASSES FOR PROTEINS (CIRCLE), REACTIONS (OCTAGON) OR PATHWAYS (STAR) CONNECTED BY DIFFERENT RELATION TYPES DISTINGUISHED BY SHADES.

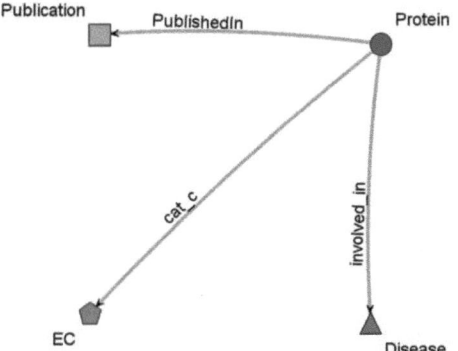

FIGURE 38 META-GRAPH UNIPROT DATABASE WITH CONCEPT CLASS FOR PROTEINS (CIRCLE) CONNECTED TO CONCEPT CLASSES OF ADDITIONAL CHARACTER, LIKE PUBLICATIONS (SQUARE) AND EC NUMBER (PENTAGON)

5 Design and implementation

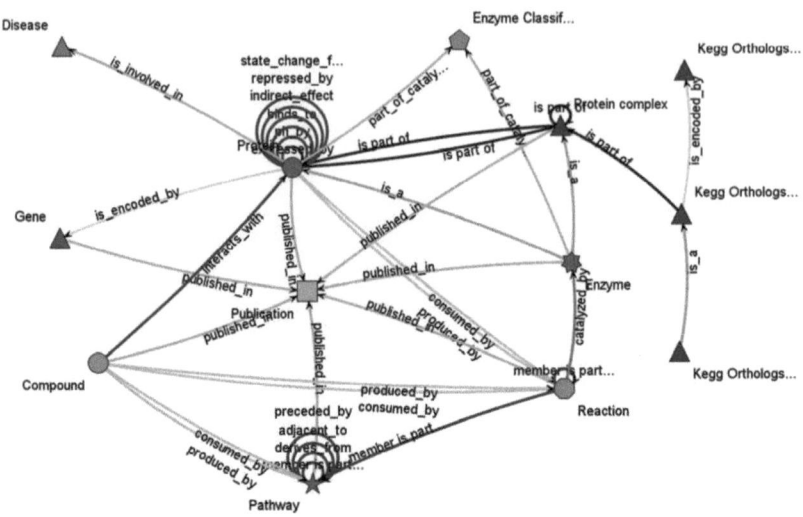

FIGURE 39 META-GRAPH OF THE COMBINED CONSENSUS DOMAIN MODEL ACROSS ALL THREE DATABASES ARACYC, KEGG AND UNIPROT

5.1.3 POPULATING THE DOMAIN MODEL

The domain model only defines what is represented in the selected domain of knowledge. Once the domain model has been created it is necessary to populate the actual instances of such a model, for example these would in the biological knowledge domain be individual genes, enzymes or reactions.

Data from several data sources are imported using data source specific parsers or generalised importers for supported exchange formats (see 9.1 "List of data formats supported by ONDEX"). For example, by using the OBO format importer, it is possible to import about 50 different ontologies (see http://obofoundry.org/), and the PSI-MI interface provides access to 8 protein interaction databases (see http://www.psidev.info/index.php?q=node/60#data), supporting this format. There are also importers for XGMML (see http://www.cs.rpi.edu/~puninj/XGMML/draft-xgmml.html) and SBML. In addition, parser for KEGG [65], Transfac [118], Transpath [119], Drastic [124], EC [70], BioCyc [66], MeSH [125] and Medline [126] exist. The flat-files of these data sources are stored locally and are read directly by the parsers. Therefore, it is not required to have a local mirror of these databases installed or that a permanent internet connection is present.

5.1.3.1 RESOLVING AND TRANSFORMING CONFLICTS

An essential part of the parsing process is to resolve and transform conflicts that exist between the schema of the originating data source and the domain model to be populated. This process generally involves the following steps:

1. Parsing of files of originating data source
2. Populating an interim data structure
3. Transforming content of interim data structure into domain model
4. Populating integration data structure with domain model

The previously presented domain model of the KEGG database (see Figure 37) can be seen as a good example of this process. Data in KEGG is distributed across several sub-databases, namely PATHWAY, GENES and LIGAND. The PATHWAY sub-database contains pathway maps distributed as KGML files (see http://www.genome.jp/kegg/xml/), whereas GENES (see http://www.genome.jp/kegg/genes.html) and LIGAND (see http://www.genome.jp/kegg/ligand.html) consist of semi-structured text files. Therefore different approaches to flat-file and XML parsing have to be deployed here. The main intention behind the KEGG domain model is to focus on the metabolic pathways represented in the database, an example KEGG pathway shown in Figure 40.

KEGG pathways simply consist of reaction steps between compounds which have corresponding enzymes assigned. This information is represented in KGML. Additional information like the chemical formula of a compound or the gene encoding an enzyme has to be parsed from the additional sub-databases GENES and LIGAND. This imposes the challenge of identifying linked entries between the different sub-databases and the parsing process has to be able to deal with missing information in each of them.

For example, as KEGG pathways are automatically derived based on orthologous relationships between genes of reference organisms and the selected species, it is possible that genes for some enzymes are not present in the selected species. In such case, no concepts of concept class Protein and Gene can be created and the Enzyme would be marked as "derived" in the integration data structure.

5 Design and implementation

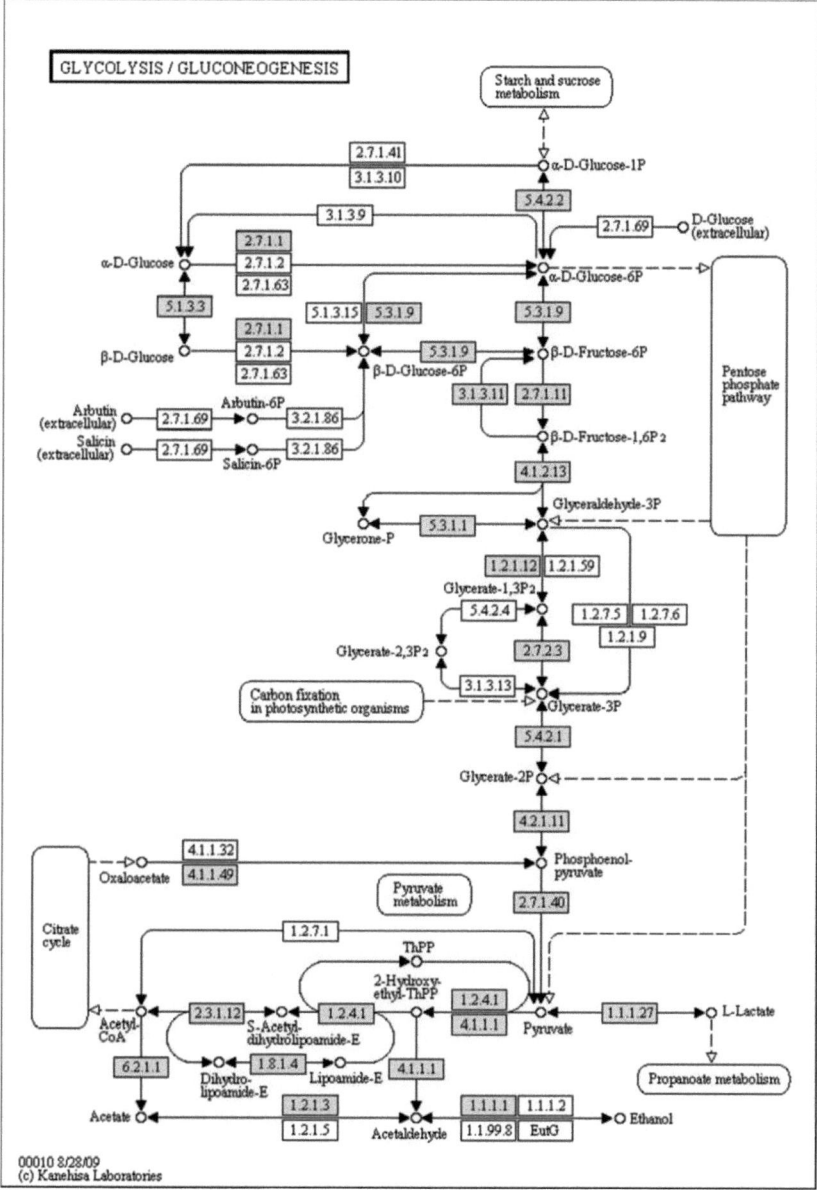

FIGURE 40 PATHWAY MAP ATH00010 FROM KEGG WITH ORGANISM SPECIFIC ENZYMES HIGHLIGHTED AND METABOLITES REPRESENTED BY SMALL WHITE CIRCLES.

5.1.3.2 STRUCTURED DATA FORMATS

Several initiatives have emerged to distribute biological data in a structured format. These include file formats like SBML [101], BioPAX [90] or PSI MI [98]. SBML, the System Biology Mark-up Language, is a computer-readable format for representing models of biological processes. It is applicable to simulations of metabolism, cell-signalling, and many other topics (see http://www.sbml.org). BioPAX Level 3 covers metabolic pathways, molecular interactions, signalling pathways (including molecular states and generics), gene regulation and genetic interactions. BioPAX Level 3 is currently under development and reviewed by pathway data source providers (see http://www.biopax.org). PSI MI, the HUPO PSI's molecular interaction format, is a community standard for the representation of protein interaction data (see http://www.psidev.info/).

Data source providers like KEGG or BioCyc slowly adapt such formats. The BioPAX wiki entry for KEGG (see http://biopaxwiki.org/cgi-bin/moin.cgi/KeggConversionSummary) reports that conversion "results in an approximately 80% pathway conversion rate, though this rate is an average over all organisms and some organisms may have higher or lower rates". Unfortunately due to the complexity of the originating data sources a 100% conversion rate is difficult to achieve. For the above KEGG example a reason given is: "All the reactions involved in a metabolic pathway are converted into BioPAX biochemicalReaction except the reactions involved in a 'multiple enzymes/multiple reactions' relation e.g. reaction 1.2.1.8 of the eco00260 pathway. The problematic reactions are excluded from the conversion because there is no way to suppress the ambiguity of which specific enzyme catalyzes which specific reaction. The controller of BioPAX catalysis uses a gene product as protein, which is an approximation since an enzyme is often represented as a complex of polypeptides in a given state."

The potential ambiguity in the interpretation of such structured formats like SBML, BioPAX or OBO makes it difficult to define a generic mapping from the structured format to a domain model for the integration data structure. Usually some configuration has to be made to cater for individual interpretations per originating data source.

Despite the development of these structured data formats efficient data integration is still an open issue where further improvements are needed. In the future, further developments are needed for ontologies and the ability to link data sets between different sources. In terms of tools, there is also a further need for development of format-independent tools as well as tools for integration of data represented in the same or different formats. [127]

5.1.3.3 DEALING WITH UNSTRUCTURED DATA

Most biological knowledge is still buried in millions of scientific articles as free text. There is currently no universal approach to extract knowledge from scientific literature. Nevertheless, several point solutions have been developed. For example, for the extraction of protein-protein interactions from scientific text [128] or functional relationships among genes [129]. Results of such extraction steps can readily be used to populate the integration data structure with a domain model relating to each application domain.

5.1.4 DATA FILTERING AND KNOWLEDGE EXTRACTION

To be able to explore large and densely connected graphs, methods for data filtering and knowledge extraction are essential in order to reduce overall complexity. As shown in Figure 33, the final stage in the data integration process is the analysis of data in client tools, which usually allow the extraction of sub-graphs according to certain criteria. These criteria are defined with respect to ONDEX metadata, for example concept classes or types of relations, the graph structure (like the degree of a node), and context information associated with concepts and relations.

A new feature in the most recent version of the ONDEX system is the support for contexts (see 4.1.8 "Context") which allow relations and concepts to be annotated or qualified with other concepts in the graph. Contexts permit a finer level of classification than would have been possible with metadata alone. For example, concepts that are components of biological pathway data sources, such as proteins or reactions belong to certain pathways and are linked to certain cellular locations. These concepts are therefore qualified by having pathways or cellular location included in their lists of contexts. It is therefore possible during knowledge extraction to restrict the results returned to the corresponding sub-graph of a pathway or cellular location and thus reduce the number of nodes and edges returned significantly.

For equivalent or related entries that were identified by the mapping methods presented in previous chapters it might be necessary to copy context information across different data sources in order to include them during the process of knowledge extraction. This is achieved using the *copycontext* transformer, which extends context annotation across newly created or existing relations. New insights may emerge by transferring context information across data sources. For example, a protein is identified as belonging to a certain pathway having a known small molecule inhibitor within one data source and is characterized as being expressed in a certain tissue by another data source. By combining the context information from these two sources it is possible to infer that a pathway may be inhibited in this particular tissue type by the inhibitor.

A special kind of transformer is the "relation collapse" transformer. This transformer processes the graph-based integration data structure, merging together concepts that satisfy the defined semantic constraints, like concept class. Hence new super concepts are created that represent and replace clusters of nodes. The redundant concepts are subsequently removed. By collapsing concepts identified to be equivalent, the number of nodes and edges in the graph can be reduced.

In addition to the two methods outlined afore which perform graph transformation and filtering (transfer of context information and collapsing equivalent concepts) several other filter and transformer methods have been implemented in the ONDEX system:

Concept class filter: Concepts of a given concept class and their corresponding relations are removed from the graph-based integration data structure.

Relation type filter: Relations of a given relation type are removed from the graph-based integration data structure. This might result in some previously connected concepts becoming unconnected.

Unconnected filter: Removes concepts (nodes) with a degree of zero from the graph-based integration data structure. Unconnected concepts do not usually have any value to the information in the graph.

All of these methods can be flexibly linked together to create a workflow for particular user applications. The resulting graphs can be exported in an ONDEX specific XML (see 4.3 "Exchanging integrated data") or RDF dialect for which generic exporters are available [7] and loaded into the ONDEX Visualisation Tool Kit (OVTK) [2] for further analysis.

5.1.5 WORKFLOWS

Each step in the process of data integration presented in Figure 33 is controlled by an ONDEX workflow enactor, which processes user-defined scripts written in XML. The workflow API accepts structured XML describing a workflow run and is able to dynamically instantiate every part of the ONDEX system in a defined order. Figure 41 shows a simple ONDEX workflow using the in-memory representation of the graph-based integration data structure (see 5.2.6 "Graph implementations and persistency"), calls the KEGG parser for species "acb" and at the end uses the OXL Export to write the results to an output file. Additionally to writing ONDEX workflows as XML, a graphical user interface (see Figure 42) has been developed by fellow student Artem Lysenko to support the user in the process of defining workflows. Figure 42 shows the same ONDEX workflow as defined by the XML file. The ONDEX workflow manager also facilitates simple enacting of the workflow itself.

5 Design and implementation

```xml
<?xml version="1.0" ?>
<Ondex xmlns:xsi="http://www.w3.org/2001/XMLSchema-instance" xsi:noNamespaceSchemaLocation="ONDEXParameters.xsd">
 - <DefaultGraph name="KEGG" type="memory">
    <Parameter name="ReplaceExisting">true</Parameter>
    <!-- default true -->
   </DefaultGraph>
 - <Parser name="kegg" datadir="/importdata/kegg">
    <Parameter name="Species">acb</Parameter>
   </Parser>
 - <Export name="oxl" datafile="kegg.xml">
    <Parameter name="GZip">true</Parameter>
   </Export>
</Ondex>
```

FIGURE 41 SIMPLE ONDEX WORKFLOW XML. FIRST STARTING WITH AN EMPTY IN-MEMORY GRAPH, FOLLOWED BY A KEGG PARSER RUN FOR SPECIES CODE "ACB" AND FINALLY EXPORT OF RESULTS TO OXL.

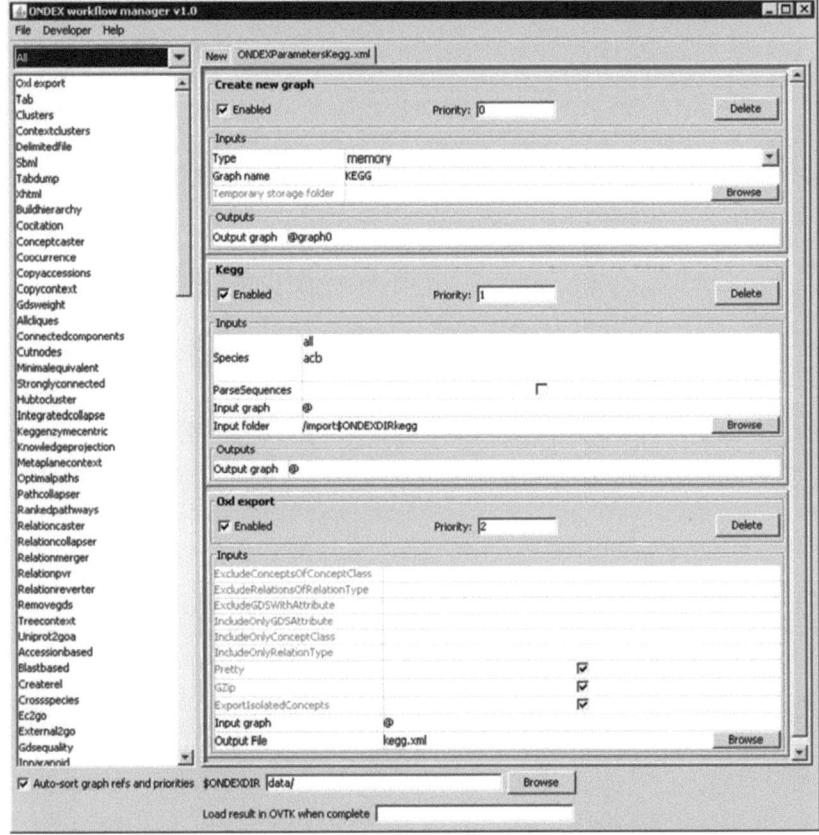

FIGURE 42 ONDEX WORKFLOW MANAGER ALLOWS SELECTING AVAILABLE WORKFLOW COMPONENTS FROM THE LEFT AND ADDING THEM TO THE CURRENT WORKFLOW ON THE RIGHT. PARAMETER FORMS CAN BE FILLED IN FOR EACH COMPONENT.

5.2 IMPLEMENTING INTEGRATION DATA STRUCTURE

The graph-based integration data structure (see 4.1 "ONDEX integration data structure", middle part of Figure 33) has been implemented as a JAVA API (application programming interface) consisting of multiple levels of functionality. Principles of object oriented design [130] have been used to increase reusability, maintainability and intuitiveness of the code. The main design criterion has been to model the JAVA API as closely as possible to the definition of the integration data structure. The UML class diagram (see 8.1.3 "Object-oriented development and UML") of the JAVA API is shown in Figure 43.

5.2.1 ENCAPSULATION

As can be seen from Figure 43 the JAVA API is provided as interfaces only (symbolised by a "i"). Interfaces are used to encapsulate discrete implementations of functionality and therefore allow for changing or alternative implementations in the future. This design feature is important for adapting the integration data structure to a variety of storage implementations, for example in-memory, databases, triple-stores etc. This JAVA API is the smallest common touch-point of all system components.

5.2.2 INHERITANCE

Interfaces as opposed to classes in JAVA allow for multiple-inheritance. This can be seen by the example of the *ONDEXConcept* and *ONDEXRelation* which both inherit method signatures from *ONDEXEntity* and *ONDEXAssociable*. Single inheritance can be seen as a form of taxonomy, for example *Unit* "is a" *MetaData* "is a" *ONDEXAssociable*. The two separate interfaces *ONDEXIterator* and *ONDEXView* are implicit linked to instances extending *ONDEXAssociable* and *ONDEXEntity* respectively via the JAVA generic type system [131].

5.2.3 ASSOCIATION

All elements of one integration data structure belong to exactly one graph representation. This is achieved using the *ONDEXAssociable* interface, which returns a unique referent for the graph representation each instance implementing this interface belongs to. Such graph representations are instances of the interface *ONDEXGraph*. This implies a cyclic association of a graph representation with itself.

The interfaces *ONDEXConcept* and *ONDEXRelation* correspond to concepts and relations of the integration data structure and therefore represent the nodes and edges in a graph representation. To distinguish them from other members of the integration data structure and emphasise the role as entities in a graph representation *ONDEXConcept* and *ONDEXRelation* inherit from *ONDEXEntity*.

Instances of *ONDEXConcept* and *ONDEXRelation* are associated with an instance of *ONDEXGraph*. Similarly all instances of *MetaData* are associated with an instance of *ONDEXGraphMetaData* which itself is associated with an instance of *ONDEXGraph*. *MetaData* is the parent interface for different attributes of the integration data structure used to add semantics to concepts and relations, for example concept class, relation type, evidence type etc.

ConceptAccession, *ConceptName* and *GDS* represent interfaces for additional values on concepts and relations to capture data in an efficient way. *ConceptAccession*, *ConceptName* and *GDS* are all associated with the corresponding instance of *ONDEXConcept*, whereas only *GDS* is associated within an instance of *ONDEXRelation*.

5.2.4 POLYMORPHISM

The interface design as shown in Figure 43 does not make use of polymorphism for method signatures. Instead polymorphism is facilitated using the factory classes *EntityFactory* and *MetaDataFactory*, which are associated with *ONDEXGraph* and *ONDEXGraphMetaData* respectively.

5 Design and implementation

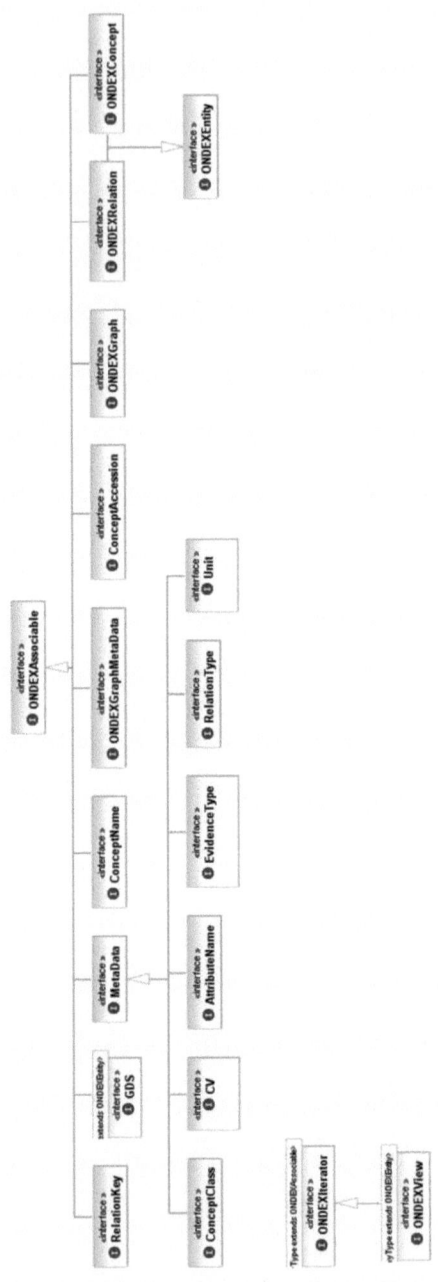

FIGURE 43 UML CLASS DIAGRAM OF JAVA API

5.2.5 ASPECT ORIENTED DEVELOPMENT

Aspects (see 8.1.4 "Aspect-oriented software development") have been used to control access at the method level described by the interfaces and therefore facilitate an efficient user permission control system which is modelled according to the established UNIX user permission model of user, group and world. Every user has a user and group id. Special ids for user root (1) and user nobody (0) exists. User and group ids are assigned to permissions for each *ONDEXEntity* and kept in an efficient Integer programming based global table structure. The Aspect system has been implemented by Jochen Weile as an extension to the pre-existing permission control system.

```
@Access(GET)
public abstract ONDEXConcept getConcept(Integer id);
```

The text box above shows an Aspect "@Access(GET)" for the *getConcept* method of the *ONDEXGraph* interface. In this example access permissions of level *GET* are necessary for the method to be executed successfully. Aspects like the one shown above get woven into the JAVA byte code at compile time and are therefore completely transparent to the user of the system. All Aspects implementations for the permission control system are contained in a separate file *AccessControl.aj*.

5.2.6 GRAPH IMPLEMENTATIONS AND PERSISTENCY

The JAVA API presented in Figure 43 allows for different implementation models. Currently two main implementations of the JAVA API exist, namely in-memory and a persistent implementation based on Oracle Berkeley DB Java Edition [120]. Both implementations share a common middle layer of abstract classes between interfaces and concrete classes. The common middle layer provides additional consistency checks for arguments passed to the methods defined in the interfaces. This three tier architecture (see Figure 44) allows for consistency across different storage implementations of the system.

5 Design and implementation

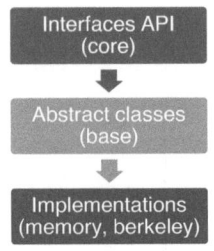

FIGURE 44 THREE TIER ARCHITECTURE OF THE ONDEX GRAPH IMPLEMENTATION STARTING WITH THE MOST GENERAL API, CONTINUING TO ABSTRACT CLASSES IMPLEMENTING SHARED FUNCTIONS AND CONCRETE IMPLEMENTATIONS AS MEMORY OR BERKELEY DB IN THE LAST TIER

The consistency checks performed by the abstract classes test attributes for null values, empty strings or missing data in general. Furthermore, behaviour constraints for the graph representation are also performed. For example, when a concept is deleted a check is made that all relations concerning this concept are deleted from the graph.

The in-memory implementation of the integration data structure makes extensive use of JAVA *Map* and *Set* interfaces. It keeps all data in the main memory of the computer during runtime. Once the system terminates, this representation of an integration data structure will lose its contents. Nevertheless the in-memory implementation is the fastest currently available.

As is common when handling large data sets, there is always a trade-off between fast data access with high memory usage and slow data access with low memory usage. To implement a low memory usage model, the Berkeley DB Java Edition has been chosen as a JAVA Object persistent storage. Other JAVA Object persistent storage systems like McObject Perst (http://www.mcobject.com/perst), DB4O (http://www.db4o.com/), Hibernate (https://www.hibernate.org/) and JDO (http://java.sun.com/jdo/) were evaluated according to the criteria of speed, usability and reliability but did not score in total as high as Berkeley DB Java Edition.

The Berkeley DB Java Edition implementation of the integration data structure uses custom methods to turn entities of the integration data structure into byte streams which are stored using the native interface layer of Berkeley DB Java Edition. This low level of direct interaction with Berkeley DB Java Edition helps to avoid significant costs imposed by higher level APIs, like the Java Collections API of Berkeley DB Java Edition (http://www.oracle.com/database/berkeley-db/je/index.html). With the Berkeley DB Java Edition implementation it is possible to load very big graphs (millions of nodes and edges) with a constant amount of memory (approximately 2GB of RAM).

5.3 GRAPH QUERYING AND INFORMATION RETRIEVAL

Retrieving data from the integration data structure is a prerequisite for further analysis and processing. Efficient and useful methods have to be provided to minimise computation time and improve ease of development for extensions to the system like new mapping methods.

5.3.1 VIEWS BASED ON SEMANTICS

The integration data structure is indexed according to metadata attributes representing the semantics on nodes and edges (see 4.1 "ONDEX integration data structure"). These indices are materialized in instances of ONDEXGraph and are returned as views (see Table 11). Views on entities are instantiations of *ONDEXView* for either *ONDEXConcept* or *ONDEXRelation* as type and therefore represent iterateable sets of concepts or relations from the integration data structure.

The default implementation of *ONDEXView* uses JAVA BitSets [132] to efficiently store members of the iterateable set. The index in the BitSet is correlated to the unique Integer ID of each concept and relation. The default implementation uses a one-to-one correlation between ID and BitSet index, i.e. a concept with ID 1234 being member of an *ONDEXView* if and only if the bit at index 1234 in the

BitSet backing this *ONDEXView* is set. Therefore *ONDEXView* does not store references to instances of *ONDEXConcept* or *ONDEXRelation* directly. It rather keeps a set tracking the existence of such instances based on Integer IDs. Other implementations of *ONDEXView* might use hash functions to keep the size of the BitSet small for very sparsely distributed unique Integer IDs.

Instances of *ONDEXConcept* and *ONDEXRelation* are only retrieved from *ONDEXGraph* by *ONDEXView* during the actual iteration process (see JAVA interface Iterable [133]). This lazy retrieval could lead to *ONDEXView* becoming desynchronised with *ONDEXGraph*, i.e. *ONDEXView* might try to return concepts which have been removed from *ONDEXGraph* meanwhile. This behaviour has to be dealt with in an implementation of *ONDEXView* by skipping over to the next valid concept or relation during iteration. As the unique Integer IDs for *ONDEXConcept* and *ONDEXRelation* are assigned incrementally, no insertion or frame shift errors can occur during the iteration process. An *ONDEXView* is an in-time snapshot of concepts or relations.

TABLE 11 ONDEX METHODS RETURNING VIEWS

Method of *ONDEXGraph*	Description
getConcepts()	Returns all concepts of the integration data structure
getConceptsOfAttributeName(AttributeName an)	Returns all concepts which have a *GDS* of a particular kind (AttributeName) assigned
getConceptsOfConceptClass(ConceptClass cc)	Returns all concepts of a particular type (ConceptClass)
getConceptsOfContext(ONDEXConcept ac)	Returns all concepts associated with a particular nested graph (context) specified by the given concept
getConceptsOfCV(CV cv)	Returns all concepts of a particular data source (CV)
getConceptsOfEvidenceType(EvidenceType et)	Returns all concepts with a particular type of provenance (EvidenceType)
getRelations()	Returns all relations of the integration data structure
getRelationsOfAttributeName(AttributeName an)	Returns all relations which have a *GDS* of a particular kind (AttributeName) assigned
getRelationsOfConcept(ONDEXConcept concept)	Returns all relations which belong to a particular concept
getRelationsOfConceptClass(ConceptClass cc)	Returns all relations which belong to concepts of a particular type (ConceptClass)
getRelationsOfContext(ONDEXConcept ac)	Returns all relations associated with a particular nested graph (context) specified by the given concept
getRelationsOfCV(CV cv)	Returns all relations which belong to concepts from a particular data source (CV)
getRelationsOfEvidenceType(EvidenceType et)	Returns all relations with a particular type of provenance (EvidenceType)
getRelationsOfRelationType(RelationType rt)	Returns all relations of a particular type (RelationType)

The main advantages of *ONDEXView* are summarised in Table 12. Especially the possibility to directly manipulate the content of *ONDEXView* using the underlying BitSet API improves the expressiveness of views. The logical functions "and", "or" and "andNot" have been efficiently implemented in *ONDEXViewFunctions* based on similar functions provided by the BitSet API. This is the foundation of querying graph entities using Boolean logic [134] for metadata attributes of nodes and edges.

5 Design and implementation

TABLE 12 ADVANTAGES OF USING ONDEXVIEW

Advantage	Reason
Low memory usage	Size of the BitSet << sum of all object references
Partial iterations are faster	Only as many entities as needed are retrieved from *ONDEXGraph*, for example retrieve first 10 concepts of a integration data structure containing 100000 concepts
Direct manipulation using BitSet	Use very efficient methods for union and intersection provided by BitSet, for example retrieve concepts of concept class A and concept class B

5.3.2 INDEXING AND QUERYING

The last section already introduced *ONDEXView* as the foundation of graph querying using Boolean logic. An example for this is shown in the box below. Query execution time is minimised by the indices according to metadata provided for graph entities. Assuming that the corresponding BitSet for a metadata entry has been pre-calculated at insertion time for entities in the graph and these BitSets are kept in hash tables enabling retrieval in constant time, then an *ONDEXView* for particular metadata can be returned in constant time by simply retrieving the exiting BitSet. If an *ONDEXView* can be returned in constant time and the implementations of "and", "or" and "andNot" of BitSet require linear time, then the total query time of a query consisting of n metadata terms on a graph of size m will take O(n*m) time. Usually n is much smaller than m (n << m), therefore the query time can be seen as linear O(m) with respect to size m of the graph.

Example of constructing a query on the integration data structure using Boolean logic for the retrieval of concepts from *ONDEXGraph* graph for data source (CV) A or B, type (ConceptClass) C and not associated with a type of GDS (AttributeName) G :

```
ONDEXViewFunctions.andNot(
    ONDEXViewFunctions.and(
        ONDEXViewFunctions.or(
            graph.getConceptsOfCV(A),
            graph.getConceptsOfCV(B)),
        graph.getConceptsOfConceptClass(C)),
    graph.getConceptsOfAttributeName(G));
```

Boolean logic does not make use of additional data fields, such as ConceptName, ConceptAccession or GDS, which are assigned to concepts and relations in the integration data structure. To cater for these extensive data processing and information retrieval requires the creation of sophisticated indices.

Such indices over additional data fields on concepts and relations are created using the Lucene indexing engine (http://lucene.apache.org) for fast full text searches. An instance of *ONDEXGraph* is passed to Lucene whenever advanced indexing is required during data processing and execution. This on-demand indexing has proven to be faster than keeping the index always up-to-date during the whole runtime lifecycle of the system. The old index is simply discarded and a new index is created from scratch. Indexing techniques involve normalisation of string case, white space and special character removing and word stemming (not for all data fields). Queries for the Lucene index are composed by a combination of search term and additional metadata restrictions as shown in Figure 45. It is possible to supply a custom Query Analyzer, search for exact term matches or use Lucene's fuzzy search engine.

5 Design and implementation

FIGURE 45 LUCENEQUERYBUILDER IMPLEMENTATION – EXCERPT OF EXPOSED METHOD SIGNATURES.

Results of a query in Lucene are wrapped as *ONDEXView*s over *ONDEXConcept* or *ONDEXRelation*. This enables seamless integration of Boolean logic query as presented earlier and fast full-text searches using the Lucene indexing engine.

6 USE CASES

This chapter first presents four use cases for ONDEX as examples for different application areas. The use cases address the topics of improving genome annotations (see 6.1), performing comparative genomics (see 6.2), building consensus metabolic networks (see 6.3) and applying graph based analysis methods to social networks (see 6.4). This is followed by an overview of the ONDEX SABR project and a brief summary of its applications (see 6.5).

6.1 IMPROVING GENOME ANNOTATIONS FOR *ARABIDOPSIS THALIANA*

Parts of this chapter have been published in the Journal of Integrative Bioinformatics in 2008, see [135].

6.1.1 MOTIVATION

Genome annotations are an essential information source used for interpretation of results from 'omics experiments. However, reliable genome annotations are not always available. A good summary of the problems inherent in the reliable annotation and re-annotation of genome sequence data has been recently provided by [136]. Salzberg describes the two key challenges as being the initial prediction of an accurate gene model from the raw genomic sequence and then the assignment of an annotation by sequence comparison with public databanks (for example GenBank). Salzberg highlights the difficulty of avoiding false gene function assignments inferred from incorrect reference database annotations. The motivation for this pilot use case for the ONDEX data integration framework is the development of an alternative approach to the assignment of reliable gene function based on reference data sources. Here, a particular focus is on alleviating the problem caused by the propagation of false functional inferences, which is a major source of the many anecdotal accounts of incorrect annotations in the sequence data sources. The proposed solution is to make explicit the interactions between the annotations from various reference data sources by using semantic data integration to create networks of links among the related concepts and entities from the reference data sources. These networks are then linked to the genes using sequence analysis methods. The annotation attached to a gene then becomes a network or graph whose structure and content can be analysed or visualised to explore the consistency of the biological information supporting the overall annotation.

Bringing biological data together coherently to extract additional meaning is a major undertaking for any Systems Biology project. The development of biological thesauri and classification systems (ontologies) continue to make it easier to link between components of different data sources. For example, by exploiting more consistent nomenclatures and using accepted lists of synonyms for biological processes and structures. This, however, only solves part of the problem of data integration because biological components can be related in many different ways. For example by taking part in a particular reaction, performing a certain function within a specific location or being part of a more complex structure. This information needs to be captured and classified accurately for it to be useful in data integration. Similarly, information about the provenance of data can be important in subsequent interpretations of any results. New types of information, such as descriptions of biological processes and pathways for metabolism and information flow, are also emerging in data sources that are valuable for linking among databases. Many of these have been created by extracting information from the scientific literature to form the basis of predictive dynamic models and simulations of biological systems. They also use complex representations that challenge traditional database systems.

An important factor, that will determine the long term success of the approach presented in this study, is selecting the best sources of functional data from among reference data sources and the best methods for integrating these data. Here quantitative comparisons between annotation steps are made and some preliminary results presented.

6.1.2 DATA INTEGRATION APPROACH

Data integration systems have been presented in Chapter 2 and for this particular application case some inspiration has been taken from Biozon (see 2.3.2). The ONDEX data integration framework provides a range of algorithms and mapping methods, suitable for identification and linking of equivalent and related data (see 4.2 "Data alignment") entries from a wide variety of data sources.

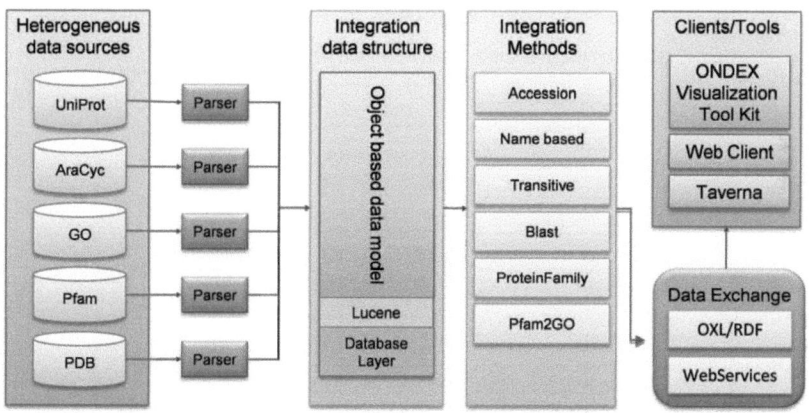

FIGURE 46 OVERVIEW OF THE ONDEX DATA INTEGRATION SYSTEM WITH COMPONENTS RELEVANT TO THIS APPLICATION CASE

In this study, ONDEX (see Figure 46) was used to integrate protein sequence data from UniProt [60] with protein structures from the Protein Data Bank (PDB) [67], protein family assignments derived from the use of the PFAM database [78], biological pathway data from AraCyc [73] and terms from the Gene Ontology (GO) [71]. These data sources were chosen as they contain some of the richest sources of protein function information and therefore are a good basis for evaluating the potential of a graph-based approach to gene annotation. The use of the GO annotation data was considered key to the later evaluation of the different approaches for mapping protein sequence to functional annotations.

6.1.3 DATA INTEGRATION EXEMPLAR

Figure 47 shows a graph of the metadata for the AraCyc database and illustrate how data integration in ONDEX can provide an elegant overview of the information captured during integration. A meta-graph (see 4.2.3.3 "Visualising results") shows all concepts classes and relation types currently in the graph; much like a database schema does for a relational database. Every entity type is represented as a concept class with all relevant relations. Further information such as cross reference accessions or synonyms is stored with the actual concepts. The meta-graph representation provides a useful high level overview of the data and helps users to understand the structure of a loaded graph.

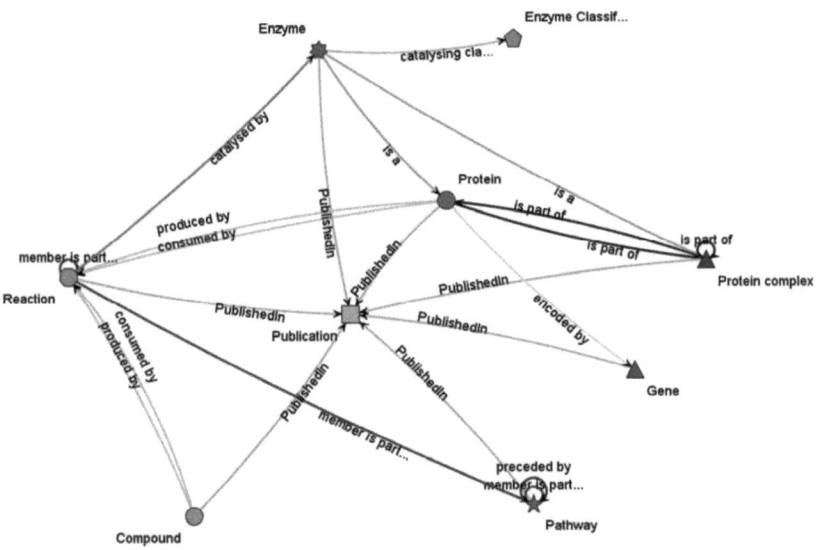

FIGURE 47 ONDEX META-GRAPH FOR THE ARACYC DATABASE, DIFFERENT CONCEPT CLASSES LIKE PATHWAY (STAR) OR REACTION (OCTAGON) CONNECTED VIA DIFFERENT RELATION TYPES LIKE "MEMBER IS PART" (DARKER SHADE)

AraCyc is a biochemical pathway database for *Arabidopsis thaliana* containing information about enzymes, proteins, reactions, compounds and genes and how they are related to each other. For each pathway component (for example protein, reaction), additional annotations such as cross database references, publication references, or in the case of enzymes, GO-terms and EC-numbers, are also available.

The ONDEX Visualisation ToolKit (OVTK) user interface uses context (see 4.1.8 "Context") to facilitate a pathway-oriented view of the integrated dataset. Figure 51Figure 55 shows an example pathway view with context. Any given pathway concept provides a context for all constituent elements of that pathway. By selecting a particular pathway from a list, all of the associated data for that pathway is displayed without the need for additional filtering. The result is similar to querying the web interface of the AraCyc database.

6.1.4 DATA INTEGRATION PIPELINE

A simple data integration workflow is presented schematically in Figure 48. Information from the AraCyc database, UniProt, and the PDB is combined with GO-terms and PFAM protein family information to provide richer annotations. Structural information from PDB was mapped to protein families; GO terms annotated to these families were used to infer protein function. The pipeline consists of 16 steps grouped into four blocks that can be separated into four distinct stages, shown by colour and explained in the following:

Step 1 to 5: Integrating AraCyc, GOA and UniProtKB

The release of AraCyc at the time of writing contained information for about 6025 proteins and a number of protein complexes. In the second pipeline step, protein sequences for all of these entries were obtained from the UniProt database.

At the time of writing, the UniProt database contained approximately 5.5 million entries, of these 349,480 were manually curated (UniProtKB/Swiss-Prot) and the remaining 5,329,119 were automatically annotated (UniProtKB/TrEMBL). Both Swiss-Prot and TrEMBL were used to create the integrated dataset.

GO annotations for the protein entries in UniProt were taken from the GOA database cross reference files. These files provide links from each protein to their manually curated GO-terms. The GOA parser creates concepts of class protein connected to concepts for GO terms.

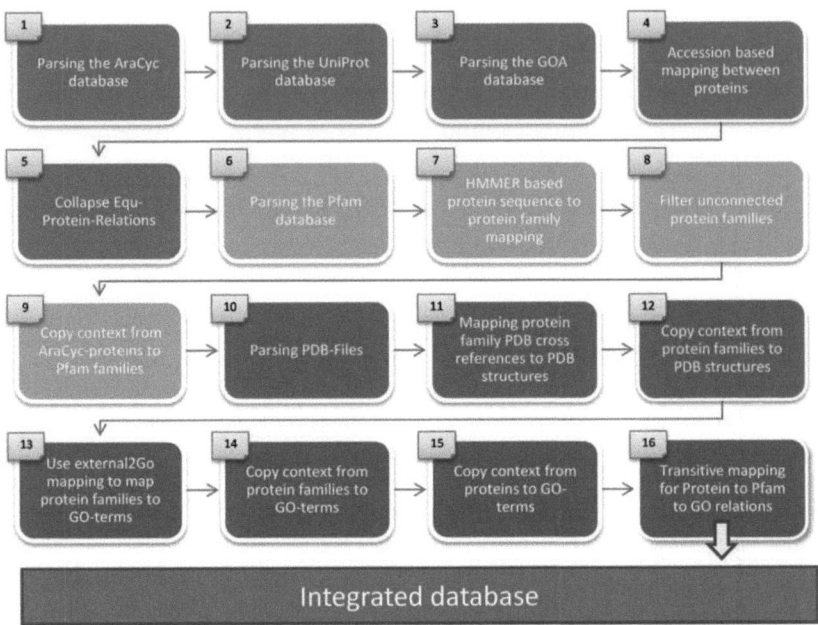

FIGURE 48 PIPELINE FOR ANNOTATING PROTEIN SEQUENCES WITH GO TERMS, PROTEIN FAMILY INFORMATION AND PDB STRUCTURES. 1-5: INTEGRATING ARACYC, GOA AND UNIPROTKB; 6-9: ADDING PFAM-FAMILY INFORMATION TO PROTEINS; 10-12: MAPPING STRUCTURAL INFORMATION; 13-16: MAPPING GO TERMS TO PROTEINS.

After the parsing process, two equivalent protein concepts may exist for each protein mentioned in AraCyc. These were mapped using accession based mapping to combine the data from AraCyc, UniProt and GOA. Where there were multiple matches in UniProt for the same entry in AraCyc, the manually curated one in UniProt was preferred. Where there were no manually curated annotations, automatic annotation was used instead. In all cases only one annotation per sequence was used in order to avoid redundancy in the test set. Afterwards the "collapse filter" (see 5.1.4 "Data filtering and knowledge extraction") was used to merge the information from these three sources into one super concept (see Figure 49).

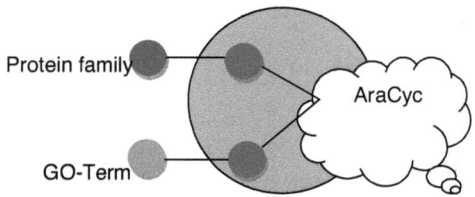

FIGURE 49 AFTER THE "COLLAPSE FILTER" WAS APPLIED THE PROTEINS (CENTRE) WERE MERGED INTO ONE CONCEPT.

Step 6 to 9: Adding PFAM-family information to proteins

PFAM is a high quality, publicly available protein family database, which is based on Hidden Markov Model profiles (HMM) and maintained by the Sanger Institute. It provides cross-references to structure information and GO annotations via GOA.

Firstly the PFAM database was parsed into an ONDEX graph representation. The *sequence2pfam* mapping (see 4.2.2.3 "Other data integration methods") was then used to map proteins to protein families based on sequence information. The *sequence2pfam* method supports three ways of mapping a protein to a PFAM protein family:

    a) Via a publicly available implementation of HMMER
    b) Using the "TimeLogic" implementation of HMMER from Active Motif, Inc.
    c) BLAST search with PFAM domain information derived from NCBI Conserved Domain Database (CDD) database

For the workflow presented here, the "TimeLogic" (http://www.timelogic.com) implementation of HMMER was used because it has a higher throughput and sensitivity compared to the BLAST approach. Protein families with no associated proteins were removed using the *unconnected* filter (see 5.1.4 "Data filtering and knowledge extraction"). Afterwards, context information was copied to relations created via the *sequence2pfam* mapping method.

Step 10 to 12: Mapping structural information

After protein family classifications were added, PDB structures associated with each protein family were assigned to related proteins by traversing the new mappings. For reasons of space-efficiency, it was not practical to incorporate the entire set of crystallographic coordinate data for each protein into the graph representation directly. Instead, PDB coordinates are loaded on demand by the integrated Jmol PDB-viewer (http://www.jmol.org) whenever the structure view is requested by the user.

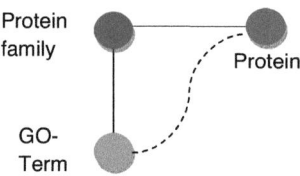

FIGURE 50 THE DASHED RELATION IS CREATED USING THE TRANSITIVE MAPPING.

Step 13 to 15: Mapping GO terms to proteins

The final part of this workflow added GO annotations to the graph. In addition to GO terms extracted from the GOA database, a second way of deriving GO annotation using the PFAM family to GO mapping file has been used. The reference file, which is provided by the GOA project, is processed by the *external2Go* mapping (see 4.2.2.3 "Other data integration methods"). Transitive mapping (see 4.2.2.3 "Other data integration methods") was used to infer protein to GO function based on protein family GO annotations (see Figure 50).

The final ONDEX integrated data graph was visualised using the OVTK. An example Pathway is displayed in Figure 51. As can be seen the Pathway (red stars) is enriched with GO-Terms (pink/orange/red circles) with PFAM families (green circles) and PDB structures (purple pentagons). Furthermore one PDB structure was selected and displayed.

6 Use cases

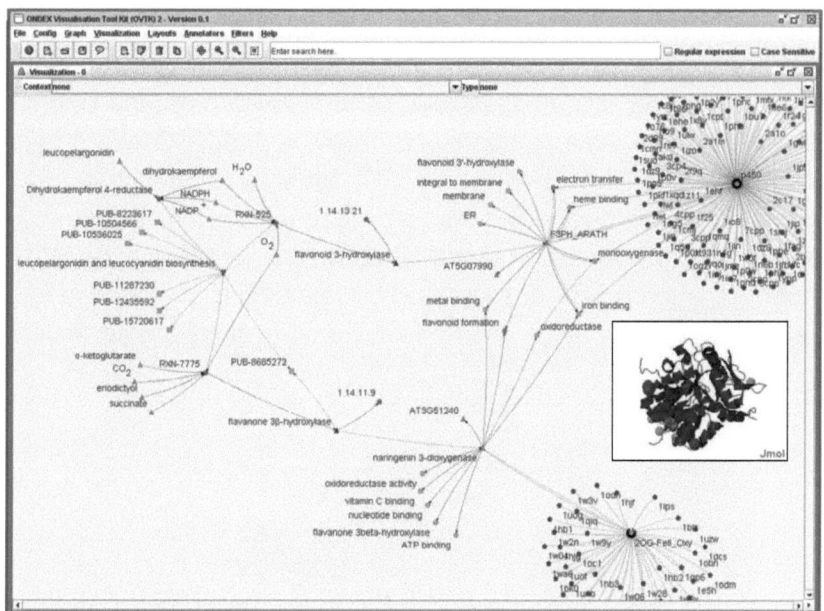

FIGURE 51 RESULTING REPRESENTATION OF AN EXTRACT OF THE "LEUCOPELARGONIDIN AND LEUCOCYANIDIN BIOSYNTHESIS". AN ARACYC PATHWAY IN OVTK WITH GO, PFAM AND PROTEIN STRUCTURE INFORMATION.

6.1.5 EVALUATION OF ANNOTATION METHODS

The motivation for using a data integration approach to sequence annotation is to improve the accuracy of automatic annotation processes. This will enable tools to be developed that can assist users and data source curators to assess the quality of a functional assignment. A quantitative measure of the accuracy of an assignment is required so that different annotation methods can be compared. This is particularly important for assessing annotations based on integrated data resources, because there will be additional uncertainties in the quality of the reference information being used and the success of the integration methods.

The first step in such an assessment is to establish a reference set against which other annotations and annotation methods can be compared. To evaluate the quality of annotation derived from sequence homology and structural classification of proteins in this study the Gene Ontology Annotation (GOA) database was used as the reference set. The reason for selecting GOA was because it provides high quality annotation of gene products from the UniProt Knowledgebase either by computationally deriving GO terms from SwissProt, InterProt, HAMAP and EC numbers or through manual curation of scientific publications. In the evaluation (Figure 52) 20458 UniProt entries for *Arabidopsis* that are annotated in GOA (~58% of entries) have been used. This relatively high coverage and the quality of annotation itself make it a very good choice for the reference dataset.

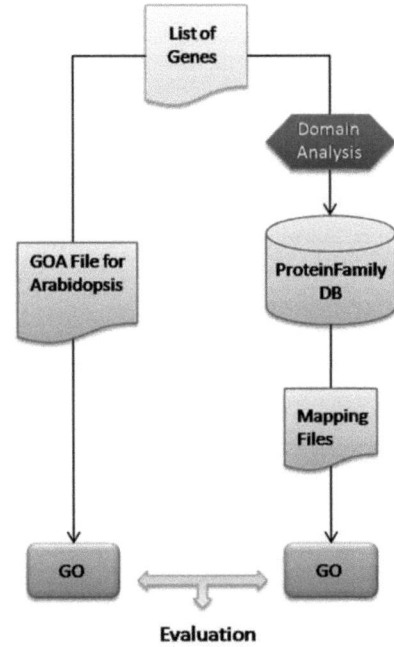

FIGURE 52 EVALUATING PFAM BASED GO MAPPINGS TO AN *ARABIDOPSIS* REFERENCE SET FROM PUBLICLY AVAILABLE GOA FILES.

The test set was compared to the reference set using the weighted harmonic mean of precision and recall (F_1), a recognised standard test for measuring performance of information retrieval methods [81].

Table 13 summarises the results of the comparison between manual GOA annotation and the predicted annotation and shows precision, recall and F_1-Score for the three GO ontology categories. The best F_1-Score was achieved for identification of biological process. The second best performance was annotating by molecular function. These results are perhaps not surprising, because most PFAM entries are constructed from functionally equivalent protein families. The performance of the PFAM annotation pipeline for both biological process and molecular function undoubtedly reflects the importance of conserved sequence and structure features in determining function. The difference in granularity between the definitions of molecular function and biological processes may also contribute to the divergent performance of predictions in these categories. By contrast, the cellular component category annotation had the lowest recall and therefore the lowest F_1-Score. This is also reasonable, because the PFAM sequence profiles are generally less able to predict cellular location (for example targeting signals) while the GO ontologies capture cellular location in some detail. Expert curators of protein data sources will also have used other information such as the original publication to assign the cellular location for the protein. The result of comparing PFAM-based annotations with the reference GOA annotations for cellular component is perhaps notable for having a greater level of success than might have been expected. These results have been published in the Journal of Integrative Bioinformatics in 2008, see [135].

TABLE 13 EVALUATION OF ANNOTATION QUALITY FOR DIFFERENT GENE ONTOLOGY CATEGORIES.

	Molecular Function	Biological Process	Cellular Component	Overall
Precision	0.67	0.82	0.86	0.73
Recall	0.43	0.58	0.26	0.45
F_1-Score	0.52	0.68	0.40	0.56

6.1.6 Discussion

In this study some preliminary research into an integrative approach for the annotation and re-annotation of genome sequence data has been presented. The first objective was to extend the range of data sources used in the data integration framework ONDEX and develop a workflow that integrates the data sources considered the most valuable for assigning protein function (i.e. curated protein sequences (UniProtKB), gene classification ontologies (GO), protein structure (PDB) and protein family assignment (PFAM)). The semantic integration of these data sources is the core of a developing platform that will be used to augment the annotation of emerging genome sequences being studied at Rothamsted Research and by collaborators. The ONDEX visualisation toolkit (OVTK) has many features that will support more detailed scrutiny of the gene and protein function assignment by making explicit the links between the different data sources and when biochemical pathway information is also available, this can also be integrated and visualised to show the broader biological context of the genes of interest including displaying protein structure information where this is available.

During this research, it was clear that at key points in the integration process, alternative mapping methods could be used to provide the links between protein sequence and GO annotations and it was therefore important to quantify the success of these methods as part of the validation process for the integration workflow. Comparing the success of these mapping methods is similar to the bigger problem of assessing the success of assigning a biological function to each protein sequence derived from the coding sequences in a new genome. The comparison between different annotation approaches is however, not straightforward and a number of different statistical methods and criteria for successful classification are used, for example hierarchical evaluation measure [137] and Fisher's false discovery rate [138]. Although complex statistical measures have some advantages, such as dealing with partially correct annotations, the classic precision, recall and F-score still remain among the most widely recognised measures of information retrieval quality because they are much more straightforward to interpret and compare than more complex tests. For this reason these common measures to explore the performance of the functional annotation methods presented in this study were selected. The results of the evaluation indicate that it is possible to use protein family categorization based on multiple sequence alignments to successfully infer biological and molecular function with a reasonable accuracy (82% and 86% respectively) by comparison with expert manual annotation. Recall depends on the actual classification made, i.e. molecular function, biological process or cellular component for the reasons detailed in Section 6.1.5.

The selection of a comprehensive reference set ("Gold Standard") has proved to be a challenging task, as it is difficult to satisfy the needs of different methods with the same reference set. Constructing a reference set from publicly available GOA files has the advantage that the information is readily available from GOA; no further annotation work is required and the resulting reference set can be used as a common benchmark since it is freely available. This comparison of GO mapping methods has highlighted the complexity of such evaluations and further research is needed to explore the issues identified in this preliminary analysis. Future developments of the ONDEX data integration system and user interfaces will extend the range of information types that can be incorporated into the annotation process and more extensive statistical methods will be developed to analyse the data and annotations. This research will also address the broader issue of assessing the contribution that data integration brings to the annotation process. This chapter, as a report of work in progress, demonstrates that the ONDEX system can be also adapted to this task with relative ease and provides a powerful platform for future genome annotation research.

6 Use cases

6.2 PREDICTION OF POTENTIAL PATHOGENICITY GENES IN *FUSARIUM GRAMINEARUM*

6.2.1 MOTIVATION

Current high-throughput DNA sequencing techniques (for example 454 Life Science sequencing) enable the creation of large amounts of novel genomic data for a large variety of organisms including important pathogen species. Adding annotations and accurate functional predictions to coding sequences identified in newly emerging high throughput genome data remains a significant challenge in bioinformatics and is a pre-requisite to the analysis of experimental data in Systems Biology. Due to the large variety of pathogenic organisms and their high diversity of molecular mechanisms, it is difficult to study every single pathogen in the greatest detail. Transferring annotations from model organisms or closely related species based on sequence homology is generally seen as essential to address this challenge.

Developing and improving fungicides and pesticides require a better understanding of the molecular mechanisms behind pathogenicity in the respective organism. The goal is to increase the specificity and reduce the dose required of fungicides and pesticides. Improved fungicides and pesticides will result in better animal and plant protection and will have less negative influence on the surrounding biological system. Identifying candidate genes and biochemical pathways is the first step in this process. In this use case, candidate genes of *Fusarium graminearum* are examined using comparative genomics techniques combined with manual curated data about pathogenicity in other organisms.

Fusarium graminearum commonly infects barley if there is rain late in the season. It has economic impact to the malting and brewing industries as well as feed barley. *Fusarium* contamination in barley can result in head blight and in extreme contaminations the barley can appear pink. *Fusarium* also infects wheat and maize where it can also cause root rot and seedling blight. The genome of this pathogen has recently been sequenced.

6.2.2 METHODS

The PHI-base [86] database (http://www.phibase.org) developed at Rothamsted Research is a unique multi-species pathogen resource because it only contains expertly curated molecular and biological information on genes proven to affect the outcome of pathogen-host interactions. This information is retrieved from the peer reviewed scientific literature. Since 2007 it also includes information on known fungicide target sites. PHI-base at the time of writing contained 1185 entries (n = 950 genes / 1185 interactions) for experimentally verified pathogenicity, virulence or effector genes from plant, fungal, insect and animal attacking fungi, Oomycetes and bacteria.

In this use case techniques of comparative genomics have been combined with data integration methods to predict pathogenicity genes in newly sequenced pathogenic organisms. The InParanoid [139] algorithm for the prediction of orthologous groups of proteins was implemented as part of the ONDEX [2, 5] data integration framework. Data integration methods for the contents of both PHI-base and genome data from pathogenic organisms were developed for ONDEX. An integrated view of these data provides a novel platform for extracting significant new insights from the comparative genome analysis, which would not otherwise have been possible.

PHI-base (see Figure 53) is made available free of charge for academic research. It can be found at http://www.phibase.org/.

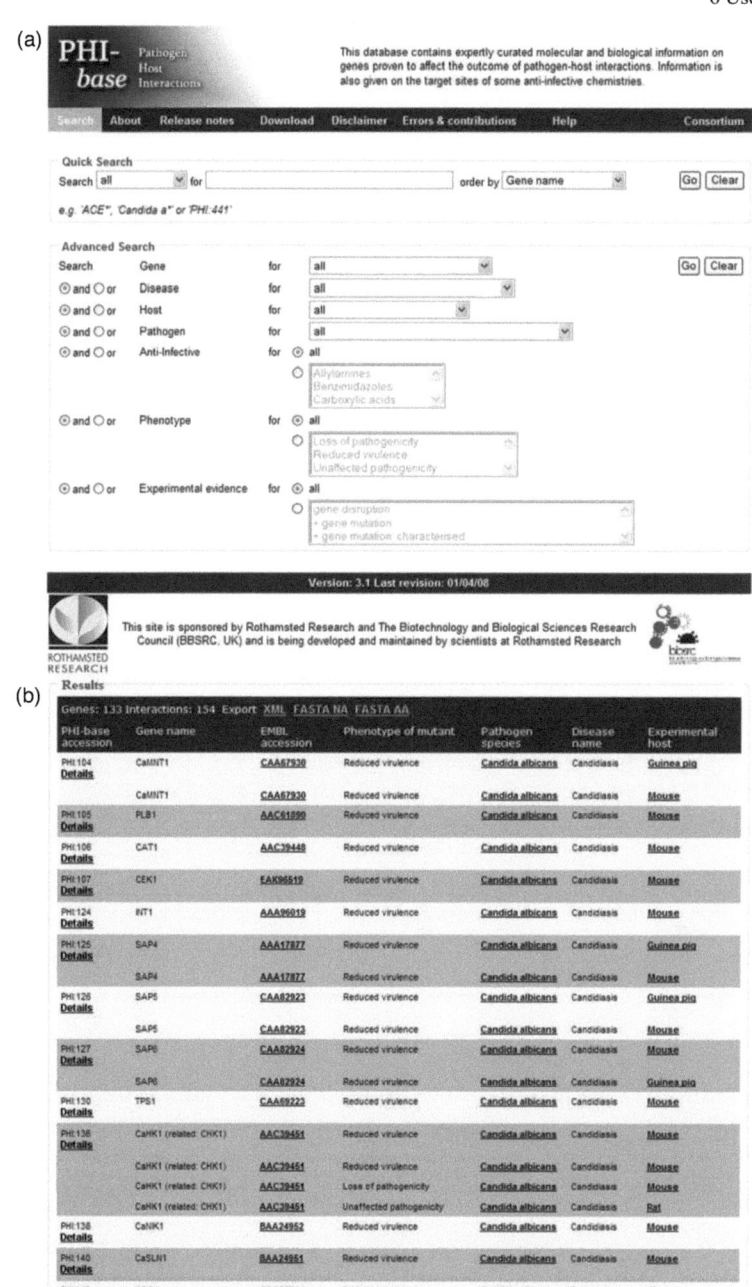

FIGURE 53 PHI-BASE, A.) FRONT-PAGE WITH QUERY MASK AND B.) RESULT PAGE FOR QUERY 'CANDIDA A*'.

6.2.3 RESULTS

A workflow (see Figure 54) has been assembled for the automated prediction of potential pathogenicity genes in combination with data extracted from PHI-base. The workflow in ONDEX consists of the following three main steps:

a) Import data into a semantically consistent data model using the graph-based integration data structure (see "4.1 ONDEX integration data structure")
b) Run implementation of the InParanoid algorithm based on sequence alignment (BLAST) mappings of the genomic data
c) Filter for significant results and export into XML (see "4.3 Exchanging integrated data").

The rich network of annotation information created through the integration workflow was visualised using the ONDEX Visualisation ToolKit (OVTK, see Figure 55) [53]. Data filtering and visualisation methods were used to identify the most plausible pathogenicity genes that were candidates for experimental validation in the laboratory. This workflow can readily be applied for any *de novo* sequenced pathogenic organism in the future.

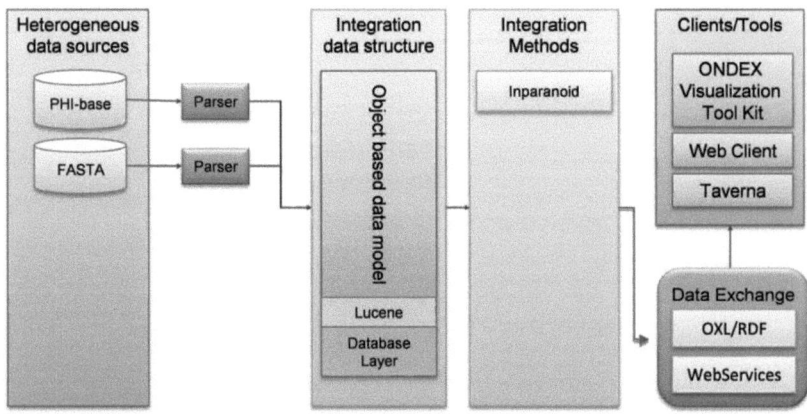

FIGURE 54 WORKFLOW IMPORTS DATA FROM PHI-BASE AND A FASTA FILE INTO THE INTEGRATION DATA STRUCTURE. THE ONDEX IMPLEMENTATION OF THE INPARANOID ALGORITHM BASED ON BLAST MAPPINGS OF THE GENOMIC DATA IS RUN. THE INTEGRATION RESULTS ARE FILTERED AND EXPORTED IN XML.

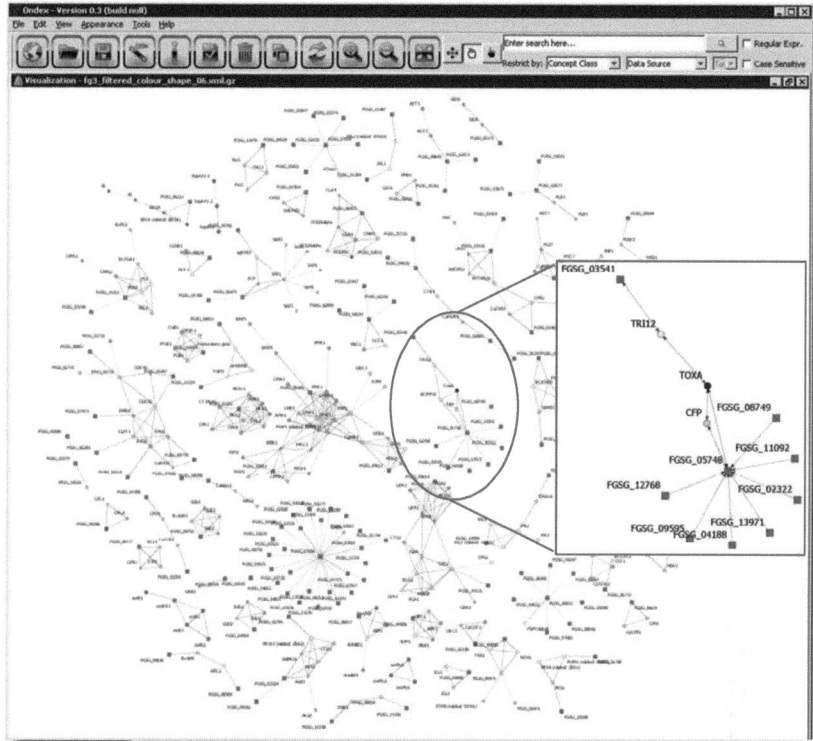

FIGURE 55 OVTK SHOWING THE RESULT OF THE INTEGRATION WORKFLOW APPLIED TO NOVEL GENES FROM *FUSARIUM GRAMINEARUM* (RED SQUARES) AND ENTRIES OF PHI-BASE (NON-SQUARE SHAPE) COLOURED BY PHENOTYPE, RELATIONS: ORTHOLOGOUS (DARKER) AND PARALOGOUS (LIGHTER)

We presented these results to our biologist colleagues, who selected interesting clusters based on their expert opinion. Figure 56 and Figure 57 are examples of selected clusters representing two extremes with respect to host response phenotypes of genes in the cluster. On the one hand Figure 56 contains genes from PHI-base of mixed host response phenotype. On the other hand Figure 57 has a more homogeneous contribution of host response phenotypes from PHI-base.

6 Use cases

In Figure 56 the *Fusarium graminearum* gene FGSG_09908 is of unknown host response phenotype (uniform shade square). All members of the depicted cluster have been identified to belong to the "Protein kinase A - regulatory subunit" family of proteins. This particular cluster does not allow for the unambiguous prediction of the pathogenicity outcome for FGSG_09908 as it contains a variety of phenotypes in other species, namely "Loss of pathogenicity", "Unaffected pathogenicity", "Reduced virulence" and "Increased virulence (Hypervirulence)". Surprisingly such ambiguous clusters seemed to be most appealing to our biologist colleagues. Subsequently FGSG_09908 has been subject to a gene knock-out experiment in *Fusarium graminearum* carried out by biologists in the Centre for Sustainable Pest and Disease Management, Rothamsted Research, UK. The gene knock-out experiment revealed the resulting host response phenotype for FGSG_09908 to be "Reduced virulence". This experimental result has been fed back into the newest PHI-base database release.

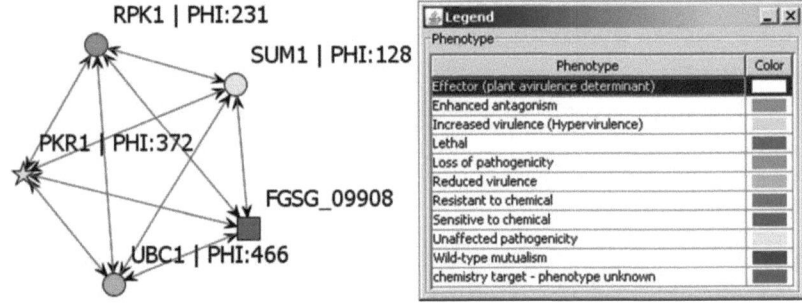

FIGURE 56 MIXED PHENOTYPE CLUSTER FOR *FUSARIUM GRAMINEARUM* WITH COLOUR LEGEND DENOTING THE HOST RESPONSE PHENOTYPE ON THE RIGHT. THE STAR SHAPE INDICATES RESPONSES FOR ANIMAL PATHOGENS AND CIRCLE SHAPE FOR PLANT PATHOGENS.

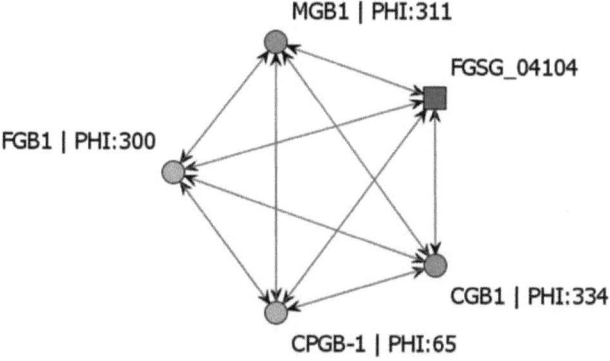

FIGURE 57 EXAMPLE OF A MORE HOMOGENEOUS CLUSTER FOR *FUSARIUM GRAMINEARUM* WITH COLOUR LEGEND AND NODE SHAPES AS ABOVE FIGURE

In Figure 57 a more educated guess on the outcome of the predictions of the pathogenicity of FGSG_04104 is possible as all cluster members are either annotated as "Loss of pathogenicity" or "Reduced virulence". The proteins involved in this cluster are part of the "G-protein beta-subunit" family of proteins. These G-proteins in *Fusarium graminearum* have recently been described in [140] to have a "Reduced virulence" effect when gene-knockout experiments were performed.

6.2.4 Discussion

Pathogens have a major impact on agricultural production and animal health. Comparative genome analysis can help biologists to better understand newly sequenced pathogen genomes. Predictions for the pathogenicity of genes can be made by cluster analysis across related species. Experimental validation of the pathogenicity of genes is an expensive and time-consuming process. Therefore prioritising targets is important.

By combining the annotated PHI-base entries with ONDEX it is possible to explore the entire contents of PHI-base as well as the gene repertoire predicted to be required for the disease causing ability of the globally important cereal attacking pathogen *Fusarium graminearum*. This has revealed which predicted proteins represent ancient conserved pathogenicity components, which are plant or animal pathogen specific and those which are single species specific. The predictions made using the presented workflow can help with target prioritisation for experimental validation.

The results for *Fusarium graminearum* have been passed to the biologists at Centre for Sustainable Pest and Disease Management (Rothamsted Research, UK) and led to a very positive response. Several predicted target genes have been prioritised and selected for experimental validation accordingly as shown in above example FGSG_09908. The outcome has been mostly in agreement with the predicted phenotype as shown in above example FGSG_04104. Currently a major publication is being prepared to present these new biological insights to the scientific community. These results will also be fed back into PHI-base to contribute to the growth of this data resource.

Eventually this data resource will get used by more and more researches in support of the development and improvement of fungicides and pesticides using the annotated information in PHI-base as target genes and processes.

6.3 CONSTRUCTING A CONSENSUS METABOLIC NETWORK FOR *ARABIDOPSIS THALIANA*

6.3.1 MOTIVATION

Data for Systems Biology is spread among hundreds of database [141]. These data sources will not always be complementary and may contain overlapping or even contradictory information. For many integrated analyses of a biological system it is desirable to have a data set that has maximum coverage with little or no redundancy. Therefore it is often necessary to construct a consensus data set from multiple sources of the same type (for example metabolic pathway resources). Effective data integration is of fundamental importance in Systems Biology research (see [13]). Related and previous work involved with constructing consensus metabolic networks, for example PROTON (see 2.2.2), has shown the importance of this research to the understanding of biological systems.

For the model plant *Arabidopsis thaliana* the two most important resources of biological pathway information are the databases KEGG [65] and AraCyc [73]. Both databases contain overlapping as well as complementary information. Structural differences in how data is represented in each of the two databases make a simple one to one mapping difficult. Therefore a user could benefit from a consensus network constructed from these two metabolic pathway databases.

Results generated during the evaluation of mapping methods (see 4.2 "Data alignment") for the mapping between the two metabolic pathway databases KEGG [65] and AraCyc [73] suggest that it should be possible to fully align these as whole metabolic pathways. To do this, it is required to identify equivalent reaction concepts, in addition to enzyme and metabolite concepts which can be mapped using the methods presented previously. A reaction concept is a placeholder for a transition of one metabolite into another catalysed by one or more enzymes. It does not have any common accession or naming conventions shared between the two databases KEGG and AraCyc. Therefore more sophisticated graph-based mapping methods need to be employed to map reaction concepts between these two data sources.

6.3.2 METHODS

A metabolic pathway usually consists of metabolites (compounds), reaction steps and proteins acting as enzymes for these reactions. Figure 58 shows the integration pipeline in ONDEX used for the construction of consensus metabolic networks. The two metabolic pathway databases KEGG and AraCyc were imported into the ONDEX integration data structure (see 4.1 "ONDEX integration data structure"). The integration methods (Accession based, Name based and StructAlign) presented and evaluated earlier (see 4.2 "Data alignment") facilitate mapping of concepts of type metabolite and enzyme between different data sources. The idea now is to go further by identifying equivalent reactions based on matching metabolites (compounds) and proteins/enzymes leading to complete metabolic pathway integration. A graph based neighbourhood search approach combined with graph pattern definition has been developed to tackle this challenge and added to the integration pipeline (Graph-pattern mapping). The results of this integration pipeline were made available in SBML format to link to other Systems Biology tools.

6 Use cases

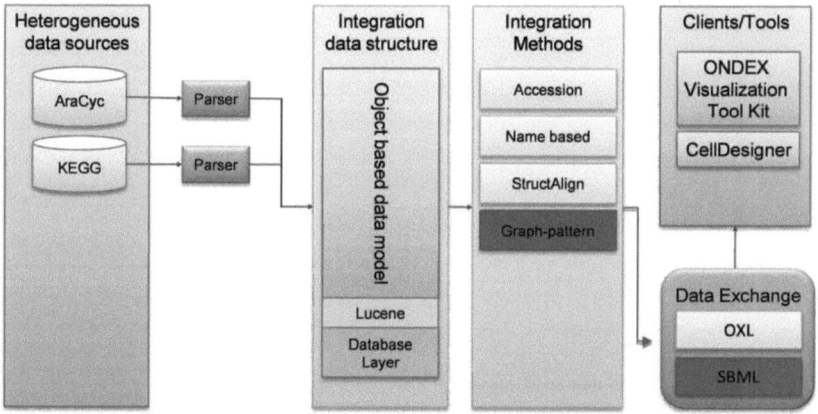

FIGURE 58 INTEGRATION PIPELINE FOR CONSENSUS METABOLIC NETWORK BETWEEN ARACYC AND KEGG, REUSING PREVIOUSLY PRESENTED INTEGRATION METHODS: ACCESSION BASED MAPPING, SYNONYM (NAME BASED) MAPPING AND STRUCTALIGN. ADDITIONALLY THE GRAPH-PATTERN MAPPING (HIGLIGHTED) AND SBML EXPORT (HIGHLIGHTED) HAVE BEEN INTRODUCED.

6.3.2.1 GRAPH-PATTERN MAPPING

Different patterns (lists) of concept classes can be used to extract paths within the integration data structure starting with a root concept class. A breadth-first search is used to traverse the graph starting at concepts of the root concept class until concepts of the leaf concept classes are reached. The concepts belonging to the leaf concept classes in each pattern are evaluated in terms of being previously identified to be equivalent. If the overlap of equivalent concepts resulting from two different paths satisfies a set threshold, then the pair of concepts at the root of each path is considered to be equivalent.

Figure 59 shows an example of how graph-pattern mapping can be applied to identify new equivalence relationships between the two different databases DB1 (blue) and DB2 (purple). Both databases contain information about proteins, enzymes and reactions. DB2 additionally contains information about enzymes forming enzyme complexes (for example protein dimers), which is missing in DB1. Reactions in DB2 are catalysed by these enzyme complexes, instead of enzymes being directly linked with reactions as in DB1.

This makes it harder to make a judgement about the relationship between the reaction concepts (3 & 7). The user of the ONDEX system is now able to define specific patterns of concept classes for each database to be used by graph-pattern mapping.

In this example the first pattern for DB1 would be Reaction -> Enzyme, whereas for DB2 the second pattern would be Reaction -> Enzyme complex -> Enzyme. For DB1 graph pattern mapping will perform a breadth-first search starting from concept 3 and traversing the graph according to the first defined pattern, which will find enzyme concept 2 as a result. For DB2 graph pattern mapping will perform a breadth-first search starting form concept 7 and traversing the graph according to the second defined pattern, which will find enzyme concept 5 as a result. The two result sets are now compared if they contain a sufficient number of concepts that had previously been identified as equivalent. As enzyme concepts 2 and 5 are marked equivalent, graph pattern mapping would now create a new relation between reaction concepts 3 and 7 as they are obviously catalysed by the same enzymes. For clarity the metabolites involved with these reactions were omitted in this example and are assumed to be the same.

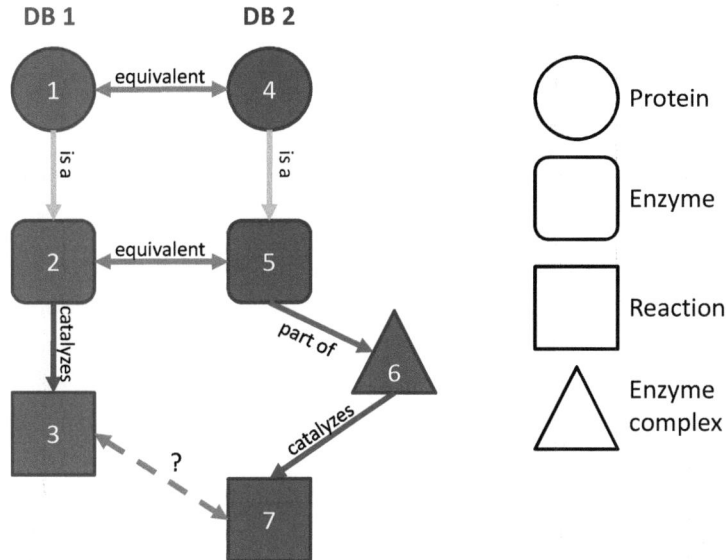

FIGURE 59 EXAMPLE FOR GRAPH-PATTERN MAPPING. CONCEPT PAIRS 1-4 AND 2-5 HAVE PREVIOUSLY BEEN IDENTIFIED AS BEING EQUIVALENT BETWEEN THE TWO DATABASES DB1 AND DB2. DB2 IS STRUCTURAL DIFFERENT FROM DB1; IT HAS ADDITIONAL ENZYME COMPLEX INFORMATION AS CONCEPT 6 (TRIANGLE). GRAPH-PATTERN MAPPING NOW TRIES TO INFER THE DASHED RELATION BETWEEN REACTION CONCEPTS 3 AND 7 (SQUARES).

6.3.2.2 SBML EXPORT

The *de facto* standard for exchanging metabolic pathways is SBML [101]. SBML is an XML based file format. Metabolites are called "species" and a simple "listOfSpecies" is used to capture all information related to metabolites. For reactions a "listOfReactions" is used to enumerate all described reactions between metabolites. It is possible for reactions to occur in different parts of the cell, so called compartments, which are described by a "listOfCompartments".

The ONDEX system has been extended by SBML Level 2 export functionality. A configuration file can be used to define the mapping from concept classes and relation types to corresponding entries in SMBL. For example, concepts of concept class compound can be mapped to SBML entries in "listOfSpecies", concepts of concept class reaction usually constitute the "listOfReactions". Only one compartment is defined by default by the ONDEX SBML export. This can be extended should the integrated data provide compartmentation information.

6.3.3 RESULTS

This analysis compared and mapped the representational structure between the KEGG and AraCyc databases. To simplify the representational schema of both data sources additional filtering has been applied which resulted in a graph representation containing the concept classes: gene, protein, protein complex, enzyme, compound and reaction. Due to the representational differences (see Figure 60) between KEGG (green) and AraCyc (red) two different patterns of concept classes are needed to identify that similar enzymes are being represented:

1. Enzyme -> Protcmplx -> Protein -> Gene
2. Enzyme -> Protein -> Gene

In AraCyc (red) a distinction between a protein and a protein complex (for example CPLXQT-6) acting as enzymes is made (pattern 1). Such distinction is not present in KEGG (green), here a simpler representation as a chain of gene (triangle) to protein (circle) to enzyme (snowflake) is used instead (pattern 2). Genes and proteins in the lower and upper part of Figure 60 had already been identified as equivalent ("equ" relationship, dark arrows) by previously described mapping methods (see 4.2 "Data alignment"). The graph pattern mapping has created new "equ" relations in the centre part of Figure 60, which has revealed that despite their representational differences the enzymatic function of these enzymes should be viewed as equivalent.

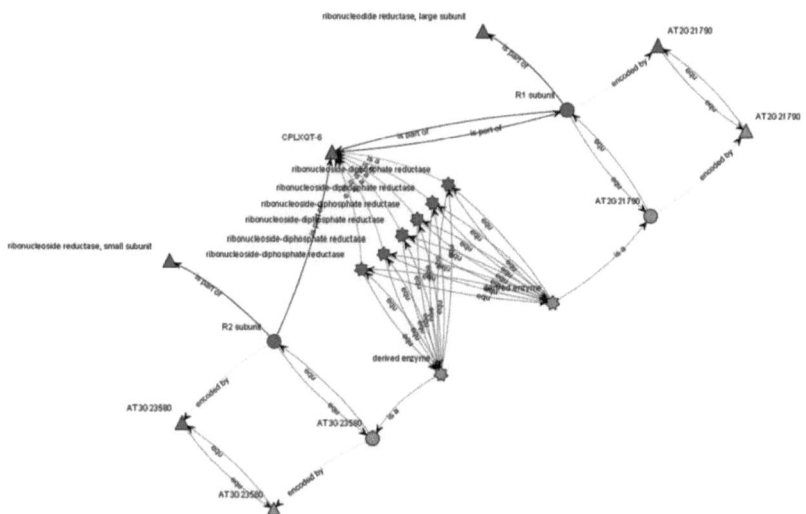

FIGURE 60 EXCERPT OF KEGG (LOWER PART) TO ARACYC (UPPER PART) MAPPING SHOWING THE STRUCTURAL DIFFERENCES BETWEEN THE TWO DATABASES IN THE MIDDLE PART (ARACYC DOES REPRESENT PROTEIN COMPLEX CPLXQT-6, KEGG DOES NOT)

Once similarity between enzymes and between compounds of KEGG and AraCyc has been derived a more simplified network can be created containing only enzymes and the compounds belonging to the reactions they are catalysing. Figure 61 presents preliminary results of the pathway integration for "Glutathione Metabolism" between AraCyc and KEGG. Triangles represent enzymes, circles represent compounds. Lower nodes are shared between AraCyc and KEGG. Upper nodes are only present in AraCyc and lower nodes are only present in KEGG.

6 Use cases

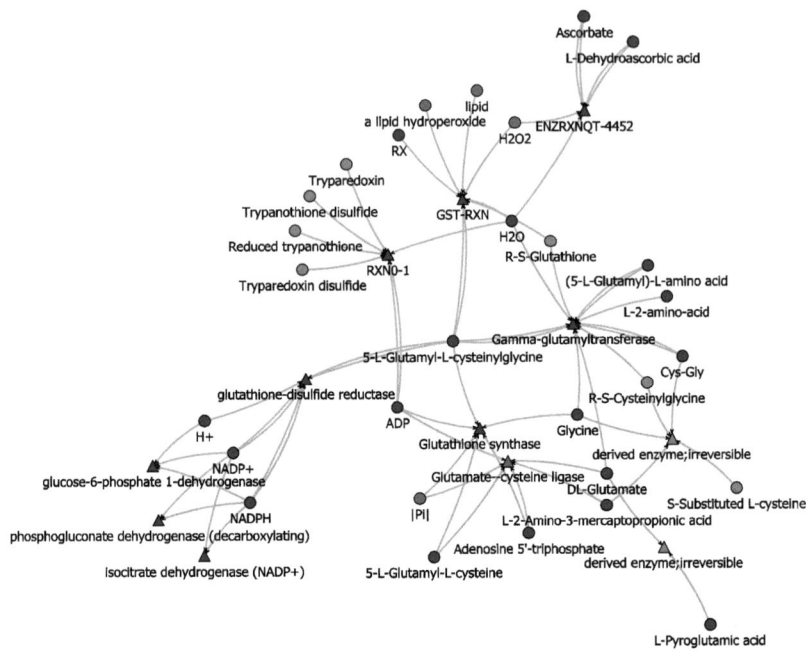

FIGURE 61 EXAMPLE: GLUTATHIONE METABOLISM IN BOTH ARACYC AND KEGG. ALL DARKER NODES ARE COMMON TO BOTH ARACYC AND KEGG. CIRCLES ARE METABOLITES; TRIANGLES REPRESENT ENZYMES IN THE INTEGRATED DATA.

The consensus network presented in Figure 61 has been exported to SBML. Figure 62 is an excerpt of the SBML file automatically created for this "Glutathione Metabolism" network. The SBML export required to define a configuration file containing the mapping from ONDEX concept classes and relation types to parts of the SBML format. This integrated network did not contain compartmentation information or kinetics on enzymes. Therefore only one default compartment is used in the SBML results and kinetic values were not set.

CellDesigner (http://www.celldesigner.org/) is one of the many tools which support SBML for loading metabolic networks. The resulting SBML has been loaded into CellDesigner (see Figure 63). Expert biological knowledge is required to assign kinetic values and starting conditions to enzymes and metabolites in the network. Once this step has been completed, the network can be used for metabolic simulation provided as part of CellDesigner.

```xml
<?xml version="1.0" encoding="UTF-8" ?>
<sbml xmlns="http://www.sbml.org/sbml/level2" level="2" version="1">
 - <notes>
    - <body xmlns="http://www.w3.org/1999/xhtml" lang="en">
        <p>Written as part of an ONDEX (url http://ondex.sourceforge.net/) export</p>
      </body>
   </notes>
 - <model id="ONDEX_Export" name="ONDEX Export">
    - <listOfCompartments>
        <compartment id="default" size="1" />
      </listOfCompartments>
    - <listOfSpecies>
        <species id="m1" name="a lipid hydroperoxide" compartment="default" initialAmount="0" boundaryCondition="true" />
        <species id="m2" name="H2O2" compartment="default" initialAmount="0" boundaryCondition="true" />
        <species id="m3" name="|Pi|" compartment="default" initialAmount="0" boundaryCondition="true" />
        <species id="m4" name="lipid" compartment="default" initialAmount="0" boundaryCondition="true" />
        <species id="m5" name="Tryparedoxin" compartment="default" initialAmount="0" boundaryCondition="true" />
        <species id="m6" name="Tryparedoxin disulfide" compartment="default" initialAmount="0" boundaryCondition="true" />
        <species id="m7" name="C3H6NO2SR" compartment="default" initialAmount="0" boundaryCondition="true" />
        <species id="m8" name="TSH" compartment="default" initialAmount="0" boundaryCondition="true" />
        <species id="m9" name="TSST" compartment="default" initialAmount="0" boundaryCondition="true" />
        <species id="m10" name="C10H16N3O6SR" compartment="default" initialAmount="0" boundaryCondition="true" />
        <species id="m11" name="C5H9N2O3SR" compartment="default" initialAmount="0" boundaryCondition="true" />
        <species id="m12" name="G" compartment="default" initialAmount="0" boundaryCondition="true" />
        <species id="m13" name="E" compartment="default" initialAmount="0" boundaryCondition="true" />
        <species id="m14" name="RX" compartment="default" initialAmount="0" boundaryCondition="true" />
        <species id="m15" name="GSH" compartment="default" initialAmount="0" boundaryCondition="true" />
        <species id="m16" name="C" compartment="default" initialAmount="0" boundaryCondition="true" />
        <species id="m17" name="C6H8O6" compartment="default" initialAmount="0" boundaryCondition="true" />
        <species id="m18" name="C5H7NO3" compartment="default" initialAmount="0" boundaryCondition="true" />
        <species id="m19" name="TPN" compartment="default" initialAmount="0" boundaryCondition="true" />
        <species id="m20" name="C8H14N2O5S" compartment="default" initialAmount="0" boundaryCondition="true" />
        <species id="m21" name="TPNH" compartment="default" initialAmount="0" boundaryCondition="true" />
        <species id="m22" name="C6H6O6" compartment="default" initialAmount="0" boundaryCondition="true" />
        <species id="m23" name="C2H4NO2R" compartment="default" initialAmount="0" boundaryCondition="true" />
        <species id="m24" name="H" compartment="default" initialAmount="0" boundaryCondition="true" />
        <species id="m25" name="H2O" compartment="default" initialAmount="0" boundaryCondition="true" />
        <species id="m26" name="ATP" compartment="default" initialAmount="0" boundaryCondition="true" />
        <species id="m27" name="ADP" compartment="default" initialAmount="0" boundaryCondition="true" />
        <species id="m28" name="C7H11N2O5R" compartment="default" initialAmount="0" boundaryCondition="true" />
        <species id="m29" name="Cys-Gly" compartment="default" initialAmount="0" boundaryCondition="true" />
      </listOfSpecies>
    - <listOfReactions>
       - <reaction id="r1" name="GGT1" reversible="false">
          - <listOfReactants>
              <speciesReference species="m25" />
              <speciesReference species="m15" />
              <speciesReference species="m28" />
              <speciesReference species="m29" />
              <speciesReference species="m10" />
              <speciesReference species="m23" />
            </listOfReactants>
```

FIGURE 62 EXCERPT SBML EXPORT OF GLUTATHIONE METABOLISM NETWORK SHOWING ALL REACTANTS ("LISTOFSPECIES") AND PARTS OF ONE REACTION LABELLED "R1".

6 Use cases

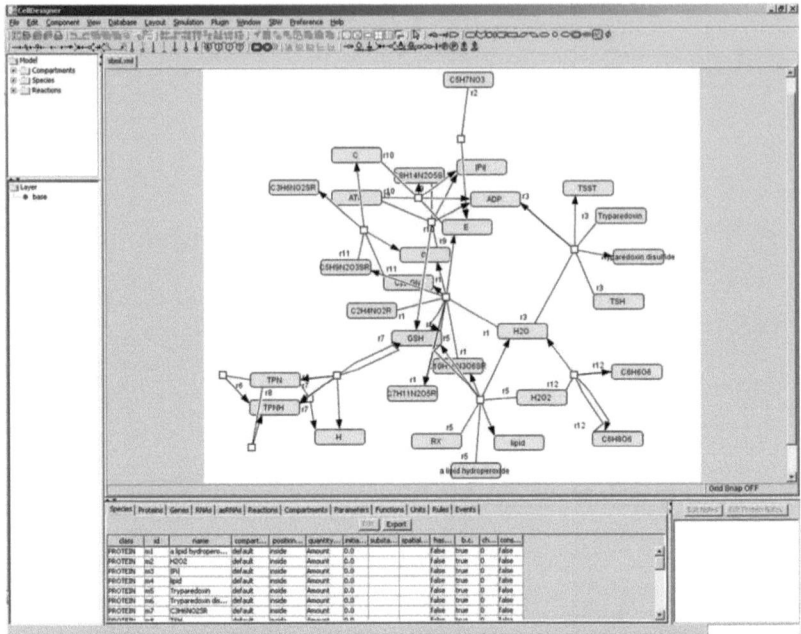

FIGURE 63 SBML LOADED INTO CELLDESIGNER. EXPERT KNOWLEDGE IS REQUIRED TO ASSIGN INITIAL CONCENTRATIONS OF METABOLITES AND CONVERSION RATES TO THE NODES AND EDGES IN THE NETWORK.

6.3.4 DISCUSSION

Differences in knowledge representation between data sources require a flexible and easy to adapt way of identifying similarities between them. This use case shows the potential of the ONDEX system to contribute to the challenges of Systems Biology by providing an integrated view across heterogeneous data sources.

Reaction concepts in the two metabolic pathway databases KEGG and AraCyc do not share a common name or identifier. None of the previously presented methods in 4.2 "Data alignment" was able to identify similarities between reaction concepts of the two databases. The presented graph-pattern mapping inferred the similarity between reactions by using previously identified similarity between enzymes and between compounds associated with reactions. This made it possible to merge similar reaction concepts. A network containing metabolites and reactions from both databases (see

Figure 61) was produced. It was possible to export this network as SBML Level 2 (see Figure 62). SBML enables results from ONDEX to be imported in other Systems Biology tools. CellDesigner (see Figure 63) was shown as an example.

The generic approach in ONDEX presented here will allow this to be used for other similar problems, for example regulatory networks. Additionally, current efforts to create a consistent overall biological ontology, like BioTop [123], could help with building consensus networks in the future. Once a consensus view across biological dataset has been established it can be used to help support generation and validation of biological hypothesis.

6.4 ANALYSIS OF SOCIAL NETWORKS

6.4.1 MOTIVATION

With the dawn of the internet as a platform for social exchange [142] supported by new Web 2.0 [143] technologies, more and more interest in the analysis of such social networks has been generated. The most prominent platforms include Facebook (http://www.facebook.com), Twitter (http://www.twitter.com) or BeBo (http://www.bebo.com). Graphs for social networks can contain millions of nodes representing the users of such platforms and millions of edges representing the relationships between users. The relationship between users in the simplest case can be "is a friend of".

This data analysis exemplar is a proof of concept of the domain independence of the mapping and analysis methods presented in previous chapters (see 4.1 "ONDEX integration data structure" and 4.2 "Data alignment"). It will highlight the challenges that have to be overcome across different domains of knowledge.

6.4.2 METHODS

The analysis of social networks requires that information distributed across the Web is captured in a computer-readable and standardized format. Such a computer readable format is the "Friend-Of-A-Friend" (FOAF) format. This format needed to be transformed into a graph representation in ONDEX. The domain of knowledge of social networks had to be reflected through the meta-data assigned to nodes and edges of the graph. A parser for the ONDEX system (see Figure 64) was created, which uses the JENA RDF framework (http://jena.sourceforge.net/) for parsing FOAF RDF and transforming it into an ONDEX graph representation. FOAF RDF files have been downloaded from different sources and parsed into the ONDEX system. Each FOAF RDF file has been marked with a unique CV (data source identifier). Existing mapping and analysis methods enable the user of the ONDEX system to consolidate and derive new knowledge from the integrated data.

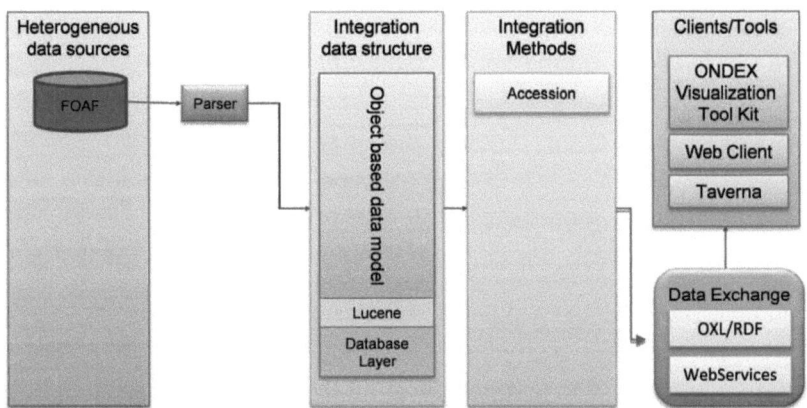

FIGURE 64 FOAF PARSER (HIGHLIGHTED) WAS ADDED TO THE ONDEX SYSTEM. EXISTING DATA INTEGRATION AND ANALYSIS METHODS COULD BE REUSED.

6.4.2.1 FRIEND OF A FRIEND (FOAF) PROJECT

The *Friend of a Friend* (FOAF) project is creating a Web of machine-readable pages describing people, the links between them and the things people do. FOAF (http://www.foaf-project.org/) defines an open, decentralized exchange format for connecting social Web sites, and the people they describe. FOAF specifications can be found at http://xmlns.com/foaf/0.1/.

Figure 66 shows an overview of available RDF classes and attributes in FOAF. These entities needed to be projected onto the structure of the integration data structure (see 4.1 "ONDEX integration data structure"), i.e. RDF classes were mapped to concept classes and the connections between them were mapped to relation types in the integration data structure.

6.4.2.2 MAPPING FOAF TO THE INTEGRATION DATA STRUCTURE

The mapping presented here is not a complete mapping of all possible RDF classes and attributes in FOAF, but represents a subset of commonly used features to enable the proof-of-concept. This approach follows three simple rules:

1. RDF classes (from FOAF) project onto concept classes in ONDEX (for example Person)
2. RDF properties (from FOAF) split into concept names (for example name, nick, firstName), concept accessions (for example mbox, mbox_sha1sum, homepage, sha1) and GDS (for example thumbnail, logo) on respective concepts in ONDEX
3. Some RDF properties (from FOAF) connect RDF classes and are transformed into relations with corresponding relation types in ONDEX (for example knows)

For this proof-of-concept the new set of meta-data for the ONDEX system was limited to concept class "Person" and the relation type "knows" (see Figure 65). The RDF properties "depiction", "mbox", "name", "publications" and "workplaceHomepage" are assigned as attributes to concepts in ONDEX according to Table 14.

6 Use cases

FIGURE 65 META-GRAPH: SUBSET OF THE FRIEND-OF-A-FRIEND (FOAF) DATA REPRESENTED IN ONDEX.

TABLE 14 SHOWS THE MAPPING FROM RDF PROPERTIES TO ATTRIBUTES ON ONDEX CONCEPTS.

RDF property	description	ONDEX equivalent
depiction	Link to a picture of a person	concept GDS
mbox	email address	concept accession
name	Full name of a person	concept name
publications	Link to a publication list	concept GDS
workplaceHomepage	Web address of workplace	concept GDS

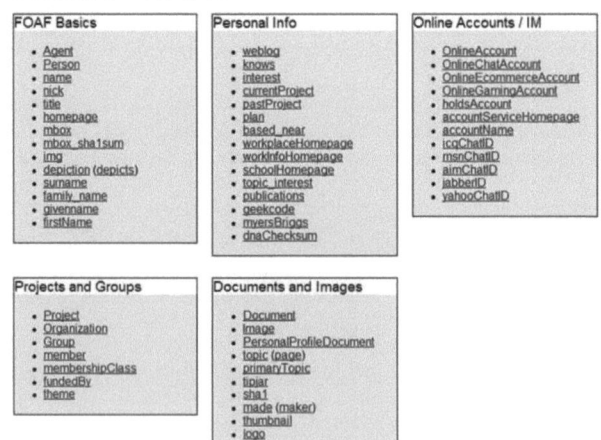

FIGURE 66 OVERVIEW OF DIFFERENT RDF CLASSES AND ATTRIBUTES IN FOAF

6.4.3 RESULTS

New domain specific meta-data for the social network domain was created containing the meta-data elements mentioned above. This meta-data was used to initialise the ONDEX system for the new domain of knowledge and five different FOAF RDF files were parsed. The result was five unconnected clusters in ONDEX (see Figure 67), one cluster for each FOAF RDF file. At the centre of each cluster is the main person for whom the FOAF RDF file was created in the first place.

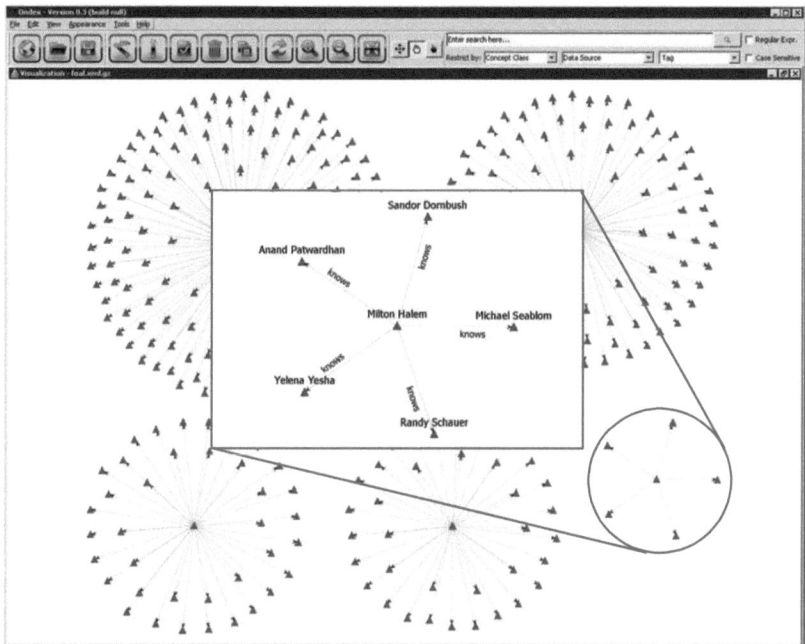

FIGURE 67 RESULT OF THE IMPORT OF FIVE FOAF RDF FILES INTO ONDEX. CONCEPTS FOR PERSON (TRIANGLE) ARE CONNECTED TO EACH OTHER BY RELATIONS OF TYPE "KNOWS" (ARROWS).

The accession based mapping method (see 4.2 "Data alignment") was used as part of the integration workflow to map concepts representing the same person (see Figure 68, new edges connecting equivalent concepts) based on the assumption that an email address (FOAF attribute mbox) is unique to one person. This resulted in a connected graph.

6 Use cases

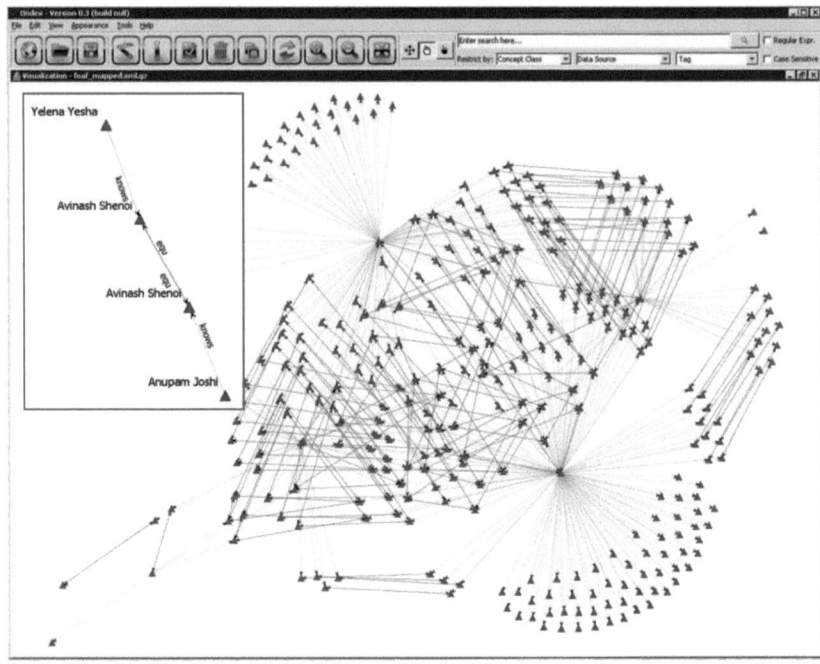

FIGURE 68 DARK ARROWS: BETWEEN CONCEPTS FOR PERSONS IDENTIFIED AS EQUIVALENT ACROSS THE PREVIOUSLY FIVE DIFFERENT CLUSTERS OF FOAF DATA USING ACCESSION BASED MAPPING. INSET: SINGLE PERSON "AVINASH SHENOI", OCCURRING TWICE. THESE OCCURRENCES WERE IDENTIFIED TO BE EQUIVALENT ("EQU").

Concepts identified to be equivalent have been collapsed using the relation collapse transformer (see 5.1.4 "Data filtering and knowledge extraction") into one, consolidating information from all imported data sources. The resulting graph (see Figure 69) contains only one type of relations "knows" as the equivalence relations have been removed by the relation collapse transformer.

6 Use cases

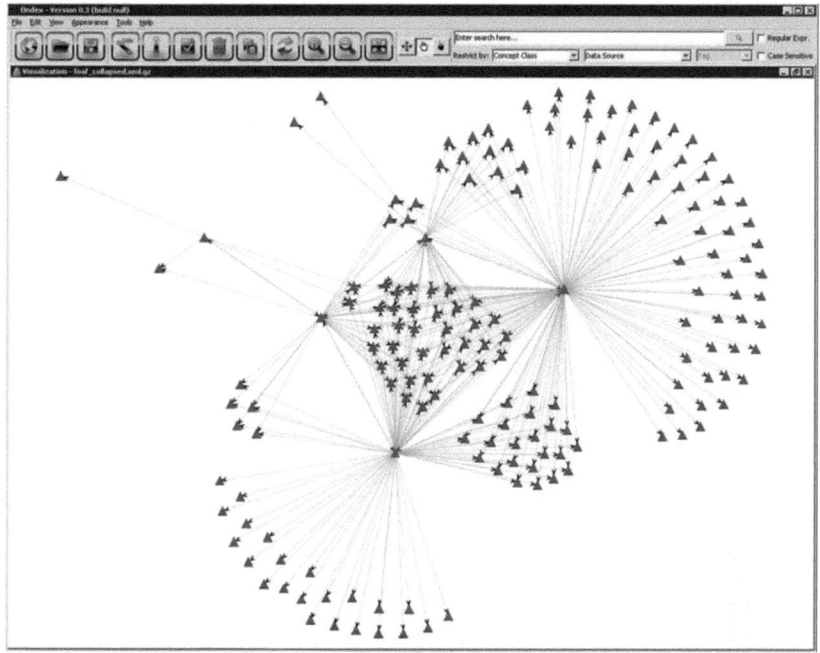

FIGURE 69 RESULT OF COLLAPSING EQUIVALENT CONCEPTS TOGETHER. ONLY ONE RELATION TYPE "KNOWS" IS PRESENT.

Different graph analysis methods provided by the ONDEX Visualisation ToolKit (OVTK) can be used to annotate graph theoretical properties to the integrated dataset, for example betweenness centrality [144, 145]. Figure 70 shows the integrated social network in the OVTK having the node size correlated with the betweenness centrality score of a node. The betweenness centrality scores are listed in the panel on the left and sorted by score. The top four scores are Tim Finin (1.0), Anupam Josh (0.49), Yelena Yesha (0.08) and Yun Peng (0.07). The names of these persons were added to the visualisation of the graph. It shows the central role of these persons in this social network. It is easy to get an impression of the number of other friends shared by these four persons just by looking at the visual representation of the graph. The cluster of friends shared between all four (in the middle of the graph) is not very large. Tim Finin and Anupam Josh share the largest number of persons, which are known between two main actors in

6 Use cases

this social network. The cluster of friends between Tim Finin and Yun Peng comes second.

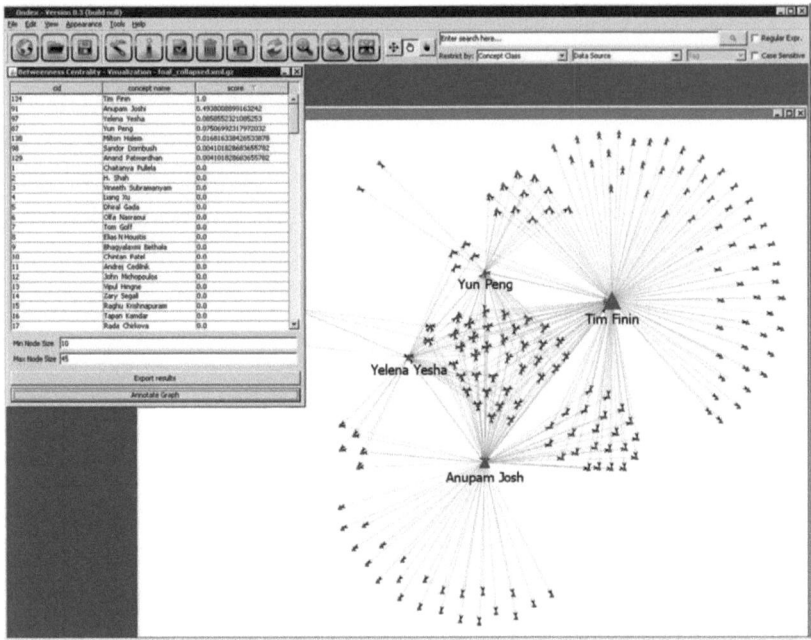

FIGURE 70 BETWEENNESS CENTRALITY APPLIED TO SOCIAL NETWORK. LEFT PANEL: BETWEENNESS CENTRALITY SCORES. SIZE OF A NODE IS PROPORTIONAL TO ITS SCORE. TOP FOUR SCORES: TIM FININ (1.0), ANUPAM JOSH (0.49), YELENA YESHA (0.08) AND YUN PENG (0.07).

6.4.4 Discussion

Despite the fact that the FOAF standard has existed since 2002, not many online social network platforms have adapted it until recently. This made locating FOAF files difficult. The Google Social Graph API (http://code.google.com/apis/socialgraph/) aims to address this problem by indexing all publicly available FOAF and XFN (http://gmpg.org/xfn/) information.

Even though the FOAF standard provides a formal framework for representing social network information, there are many inconsistencies in how the standard gets used. For example, FOAF files differed in number of tags used and information associated with tags. Exact identification of a person using only the name tag proved not feasible. Instead the email address of a person whenever provided was more useful.

Writing of the new parser for the FOAF format proved easy using the JENA RDF framework. All that was necessary to add this parser to ONDEX was to introduce a new set of meta-data as described in 6.4.2.2. Existing mapping methods, for example accession based mapping could readily be used to identify equivalence relationships within the social network data. Analysis and visualisation capabilities of ONDEX were instantly available to be applied to this new kind of data.

Information provided in form of FOAF graphs can be used in many different ways, for example augment e-mail filtering by prioritizing mails from trusted colleagues, provide assistance to new entrants in a community or locate people with interests similar to yours. Merging information, for example multiple email addresses of one person on collapsed concepts can be used by social network organisations to spot inconsistency or complete missing information across data sources. The analysis of FOAF graphs can help to reveal implicit knowledge about the connections between people, for example if they share the same friends, have other friends, the degree of connection or whether or not some people form networking hubs.

Betweenness centrality was used in 6.4.3 to identify networking hubs in the integrated social network, see Figure 70. Visualisation of the network data helped understand the structure of the network. It became easy to spot persons who are main actors in this social community according to number of friends and their connections with each other. This kind of analysis is very similar to analysis of metabolic, protein interaction or gene regulatory networks found in the life sciences. In these networks, the relationships between biological entities are examined in a similar way as has been shown for social networks. For example hubs determined by betweenness centrality measures in metabolic networks represent essential components of the metabolism of an organism. In summary, the same algorithms for deriving new insights based on graph structure can be applied to both domains of knowledge, social and biological networks.

6.5 ONDEX SABR PROJECT AND ITS APPLICATIONS

In 2007 the ONDEX project was awarded £2.7M by the Biotechnology and Biological Sciences Research Council (BBSRC) under the SABR Systems Biology initiative to create an e-tool project for supporting Systems Biology research. The ONDEX SABR project started on April 1st 2008 and lasts for 3 years. It involves three partner institutions: Rothamsted Research, Manchester University and Newcastle University with the principal technical development driven by Rothamsted. The aim of the ONDEX SABR project is to develop a robust open-source software system for integrating life science data by extending the ONDEX data integration platform. Three demonstrator applications of data integration are being developed in collaboration with BBSRC Systems Biology centres at the partner institutions which will: identify new genetic and molecular targets to improve bioenergy crops (Rothamsted, see 6.5.1.1); integrate different yeast metabolome models (Manchester, MCISB, see 6.5.1.2) and support studies of telomere function relating to ageing research in yeast (Newcastle, CISBAN, see 6.5.1.3).

In addition to the three demonstrator applications of the ONDEX SABR project, some further success stories of addressing biological problems using ONDEX are also briefly described. These use cases are part of past or ongoing collaborations with other institutions or departments. They all exploit the work in this thesis, but have been carried out by other users of ONDEX. Current use cases for ONDEX supported by the ONDEX SABR project are summarized in Table 15.

6 Use cases

TABLE 15 SUMMARY OF CURRENT USE CASES OF ONDEX SABR PROJECT

Title	Organisation	Collaborators & users
Identifying new genetic and molecular targets to improve bioenergy crops (ONDEX SABR)	Rothamsted Centre for Bioenergy and Climate Change	Part of the BBSRC Sustainable Bioenergy Centre (http://www.bsbec.bbsrc.ac.uk/)
Integration, augmentation and validation of yeast metabolome models (ONDEX SABR)	Manchester University	Manchester Centre for Integrative Systems Biology (http://www.mcisb.org/)
Supporting research into the role of telomere function in ageing (ONDEX SABR)	Newcastle University	Centre for Integrated Systems Biology of Ageing & Nutrition (http://www.ncl.ac.uk/cisban/)
Modelling processes for fruit ripening and flavour development in Tomato	Syngenta & Imperial College	Syngenta University Innovation Centre (http://www3.imperial.ac.uk/syngenta-uic)
Differences in responses to carcinogenic substances in human, rat and mouse	Syngenta & Imperial College	same as above
Plant Responses to Environmental STress in Arabidopsis (PRESTA)	Warwick University	Warwick Horticultural Research Institute (HRI) (http://www2.warwick.ac.uk/fac/sci/whri/)

6.5.1.1 IDENTIFYING NEW GENETIC AND MOLECULAR TARGETS TO IMPROVE BIOENERGY CROPS

Biomass from fast-growing trees and grasses is a sustainable source of renewable energy (http://www.bsbec.bbsrc.ac.uk/programmes/perennial-bioenergy-crops.html). In collaboration with the Centre for Bioenergy and Climate Change (Rothamsted Research, UK) ONDEX has been used to support the identification of candidate genes related to biomass increase in willow (*Salix* spp.). Data sources relevant to the comparative genome analysis of willow genetic information have been assembled and new parsers have been developed to integrate these datasets into the ONDEX system. An important focus has been on assembling a number of data resources relating to the control of plant architecture by plant hormones. Using the integrated dataset biologists have been able to identify interesting QTLs in willow based on their synteny with poplar (*Populus* spp.) genes annotated to be involved in plant growth.

6.5.1.2 INTEGRATION, AUGMENTATION AND VALIDATION OF YEAST METABOLOME MODELS

Knowledge about all biological processes even in simple organisms like yeast (*Saccharomyces cerevisiae*) is still incomplete and contains errors. Certain errors in the data can be spotted computationally. For example, some reactions consume small molecules that appear without being imported into the cell or being generated by other reactions. Similarly, some reactions produce small molecules that are neither consumed by other reactions nor exported from the cell. The goal is to find the missing reactions. Some of this knowledge does not yet exist, but much can be found in various data sources and especially within published literature. ONDEX has been used to identify missing reactions and to suggest possible fillers based on literature searches. One particularly important area for improvement of the dataset has been lipid metabolism, which is a much under-represented component of any of the existing metabolic data source. This is particularly important for the development of methods to produce bioplastics and biofuels. Where possible the missing information has been culled from the literature using ONDEX and is currently being incorporated into the latest SBML version of the Yeast Jamboree Network [146] at the Manchester University, UK.

6.5.1.3 SUPPORTING RESEARCH INTO THE ROLE OF TELOMERE FUNCTION IN AGEING

Studies of telomere function in *Saccharomyces cerevisiae* undertaken at Newcastle University (UK) have the aim to reveal new insights in the ageing process. ONDEX has been used to assemble an integrated data set of protein / gene interactions and to automate the prediction of genetic interactions from this dataset. The dataset is in use by PhD students and research assistants to manually explore the basis of genetic interactions involved in telomere function in yeast.

6.5.1.4 MODELLING PROCESSES FOR FRUIT RIPENING AND FLAVOUR DEVELOPMENT IN TOMATO

Two separate use cases are being carried out together with the Syngenta University Innovation Centre (UIC) at Imperial College (London, UK). ONDEX has been used to develop integrated datasets from public and proprietary data sources and generating Prolog forms of the data for use in machine-learning experiments being conducted in the UIC. The goal of the first use case is to better understand the molecular mechanisms underpinning fruit ripening and flavour development in Tomato. A time series experiment of different ripening stages of tomato has been conducted at Syngenta. Transcriptomics and metabolomics profiles have been assembled and are available to be included into the background knowledge for the machine learning process.

6.5.1.5 DIFFERENCES IN RESPONSES TO CARCINOGENIC SUBSTANCES IN HUMAN, RAT AND MOUSE

The second use case with Syngenta and Imperial College (London, UK) concerns identifying the differences in response to carcinogenic substances and their influence on Thioredoxin in human, rat and mouse. Thioredoxin plays an important role in cellular processes via reduction/oxidation (redox regulation) [147]. Gene expression data as well as metabolomics data is also available as time series data. Thus the methods for analysis of the metabolic networks across the different organisms share similarity and combine data integration in ONDEX with inductive logic programming [148] performed at Imperial College.

6.5.1.6 PLANT RESPONSES TO ENVIRONMENTAL STRESS IN ARABIDOPSIS (PRESTA)

In the PRESTA (Plant Responses to Environmental Stress in Arabidopsis) project at Warwick University (UK), ONDEX is being used to help reveal the gene networks implicated in biotic stress responses as part of an integrated analysis of high resolution time series expression profiles from *Arabidopsis* leaves infected with *Botryris cineraria*. Using ONDEX, a knowledge base across different data source concerning the model plant *Arabidopsis thaliana* and several pathogenic organisms has been assembles. This knowledge base was enriched using text-mining methods to associate plant and pathogenic organism proteins with an ontology of plant stress factors.

6 Use cases

7 CONCLUSION AND OUTLOOK

7.1 SUMMARY

The problem of data integration in the life sciences is well known, several approaches which address aspects already exist. A few such systems were presented in Chapter 2, namely "SEMEDA", "PROTON", "Visual Knowledge and BioCAD", "Biozon", "BNDB / BN++", "STRING" and "NeAT". This selection highlights the diversity found in tools for data integration in the life sciences. Using the review of software tools for data integration a list of challenges was identified (see Table 1). These were presented as requirements for the ONDEX system in Chapter 3. How the challenges have been addressed is discussed in Section 7.3 "Addressing the challenges".

Biological data is often organised as networks or graphs (for example protein interaction networks, metabolic pathway maps, gene regulatory maps). In Section 4.1 a graph-based approach to data integration was presented. It aims to provide both semantic flexibility, for example to be able to cope with changing knowledge, and integrity, for example dealing with errors in the data. The way how semantics or meaning can be associated with elements in the graph is presented. This was compared with characteristics of semantic networks and conceptual graphs (see 8.2 "Knowledge representation") to highlight similarities and differences. A formal definition of the ONDEX integration data structure was given at the end of Section 4.1.

Data integration has to solve the two problems of syntactic and semantic heterogeneity [57]. Among data sources syntactic heterogeneities include different data exchange formats, spelling mistakes in gene names etc. Semantic problems exist when the same entities are named differently (for example multiple gene names for one and the same gene) or different entities are named the same (for example "ear" can be the human ear or a part of a wheat plant). To overcome syntactic and semantic heterogeneity in data sources, modelling of knowledge has to be adaptable to the

7 Conclusion and outlook

respective domain of knowledge. To make knowledge explicit is a difficult task. The current model for biological knowledge employed in ONDEX requires an agreement among contributors and users of the ONDEX system and will be inspired by the biological application cases, see 5.1.2 "Formulating a consensus domain model in biology". Once a model of knowledge has been created it is necessary to fill this model with actual knowledge. In biology, for example relevant knowledge could be instances of actual genes, enzymes or reactions.

Following the import of different data sources into the graph-based integration data structure of the ONDEX system, commonalities between different data sources have to be identified as the second step. This has been achieved by using mapping methods presented in 4.2 "Data alignment". A mapping method creates a new relationship between data entries which have been identified to be equivalent or similar according to a certain criteria. A quantitative evaluation of three mapping methods, namely *Accession based mapping*, *Synonym mapping* and an algorithm called *StructAlign* was presented. All mapping methods use synonyms or accession codes of data entries in different ways to identify equivalences between them. The approach utilised in the first two listed mapping methods is easier to comprehend. The novel *StructAlign* algorithm, however, combines both information retrieval approaches and graph-based analysis and is therefore described in more detail. The three algorithms were evaluated in two representative scenarios: the integration and analysis of metabolic pathways, and the mapping of equivalent elements in ontologies and nomenclatures. The evaluation of these methods showed that a graph-based algorithm (*StructAlign*) and also simpler mapping methods through synonyms can perform as well as using accession codes only for identifying equivalences between entries. The combination of all three mapping methods yielded the most complete projection between different data sources. A particular challenge in the presented evaluation had been to identify suitable "gold standard" or reference data sets against which to assess the success of the algorithms developed.

7 Conclusion and outlook

The results presented are therefore not definitive, but represent the best statistical comparison that could be achieved in these circumstances.

The third step of data integration in ONDEX is to communicate integrated datasets to scientific peers or to load them into other tools for data analysis. For this task it has been necessary to define an appropriate transport or exchange format [7]. Section 4.3 "Exchanging integrated data" begins by introducing the requirements for the exchange of integrated datasets and proceeds to discuss and compare existing biological data exchange formats like BioPAX, MAGE-ML, PSI-MI and SBML. It was concluded that none of the outlined exchange formats sufficiently met the needs of data integration in ONDEX. Requirements for exchange of integrated datasets were described and gave rise to specification of the OXL exchange format. Its usability for data exchange between different components of the ONDEX system and external applications was demonstrated by using the PHI-base database, which supports OXL natively, as an example. Finally the various other applications of OXL were discussed, and an outlook on extension of the format and plans to improve OXL's supporting tools was given.

As data integration has to deal with very large amounts of data it is of utmost importance to provide an efficient implementation of the graph-based data structure for data integration is described in Chapter 5. The integration data structure was implemented using object-oriented programming techniques in a three tier approach. The top tier is the API consisting of a set of interfaces in JAVA. Abstract classes implementing this API in the middle tier provide consistency checks for the data passed as arguments to the methods. The bottom tier holds the data of the graph-based data structure. Currently, it makes use of the Berkley DB Java edition for data storage on the hard drive and the Lucene indexing engine to enable fast full text searches within the integrated data. The API has been designed for efficient manipulation and querying of the integration data structure. For example, concepts and relations in the integration data structure can be retrieved according to their

7 Conclusion and outlook

semantics (like concepts per concept class or relations per relation type). The results are returned as ONDEXViews. ONDEXViews are based on highly efficient BitSets and can be combined using logical operations like AND, OR and NOT in a query. With such a configuration it is possible to import, index and search combinations of data sources that could result in millions of entries. The integration process in the ONDEX system can be controlled via workflows.

To show the usability of the principles and methods implemented in ONDEX several use cases have been described in the last part of this thesis starting with a comprehensive example of addressing the challenges of how to annotate novel data generated from new high-throughput genome sequencing as a pilot project in 6.1 "Improving genome annotations for *Arabidopsis thaliana*". Here an alternative approach for assigning reliable functional information to genes based on well-known data sources is presented. This identified relationships of the novel data with reference data from well-known data sources using gene and protein sequence analysis methods. The new relationships were made explicit by creating relations between genomics data from *Arabidopsis thaliana* and databases entries of UniProt, AraCyc, GO, PFAM and PDB. The functional information associated with all entries of the resulting datasets can be interpreted as a network or graph whose structure and content can be analysed or visualised to explore the consistency of biological information supporting the overall information assignment to genes. A statistical comparison of the information assignment steps was made and some preliminary results were presented. In this study, the ONDEX system was used to integrate protein sequence data from UniProt database with protein structures from the Protein Data Bank (PDB), protein family assignments derived from the use of the PFAM database, biological pathway data from AraCyc database and classification information from the Gene Ontology (GO) database.

7 Conclusion and outlook

In Section 6.2 "Prediction of potential pathogenicity genes" techniques of comparative genetic sequence analysis have been combined with data integration methods to predict how genes in newly sequenced pathogenic organisms might contribute to the pathogenicity of these organisms. Data has been imported into ONDEX from both the pathogen-host-interaction database (PHI-base) and genetic sequence information from pathogenic organisms. The integrated datasets provided a novel platform for extracting significant new biological insights about pathogenicity and the genes involved, which would not otherwise have been possible. This analysis helped biologists to better understand newly sequenced pathogenic organisms. Predictions about the influence of genes on pathogenicity of these organisms were made using graph-based analysis across related species.

Section 6.3 "Constructing a consensus metabolic network for *Arabidopsis thaliana*" showed how the representation of information between the two databases KEGG and AraCyc could be unified. A flexible and easy to adapt way of identifying similarities between data entries despite differences in representational structure of the database has been presented. A unified metabolic network view of KEGG and AraCyc was created. The results were exported into SBML. SBML is supported by a large number of tools for Systems Biology. CellDesigner was used to illustrate how data from ONDEX can be exchanged with other Systems Biology tools via SBML. CellDesigner enables the simulation of consensus metabolic networks created by ONDEX. This use case shows the possibilities of the ONDEX system in combination with other Systems Biology tools to help biologists understand biological systems better by providing an integrated view across different data sources.

The final use case 6.4 "Analysis of social networks" was a proof of concept to illustrate the generality of the mapping and analysis methods presented in the previous chapters. It shows the steps involved when other than biological knowledge is concerned. A new domain model for knowledge about social networks and a data importer for the friend-of-a-friend (FOAF) format were created.

7 Conclusion and outlook

Existing mapping and graph analysis methods of ONDEX could be reused. Using betweenness centrality highlighted the importance of certain persons in the social network. Visualisation of the social network provided an intuitive way of understanding the relationships contained in the data. Such analysis of FOAF graphs can help to reveal implicit knowledge about connections between people. This analysis was compared to analysis of metabolic, protein interaction or gene regulatory networks found in biology. It was concluded that the same algorithms for deriving new insights based on graph structure can be applied to both biological and social network knowledge.

Chapter 6 ends with presenting the ONDEX SABR project. An overview of current usage of ONDEX within the ONDEX SABR project (see 6.5 "ONDEX SABR project and its applications") is given and each ONDEX SABR use case is briefly described.

7.2 DESIGN DECISIONS

Existing approaches to data integration can be characterised as being based on principles such as "Link integration and hypertext navigation", "Data warehouses", "View integration and mediator systems", "Workflows" and "Mashups" (see 2.1 "Principles of data integration"). The ONDEX system presented in this thesis shares most similarity with a data warehouse combined with aspects of workflows and has been designed in this way for the following reasons:

Data source availability

Few biological data sources are readily available as web-service suitable for federated access or use in Mashups. Most are only present as flat-files or database dumps. Data warehouses can make use of both: data from web-services and flat-files or database dumps using special engineered parsers or importers.

Changes in databases

Databases change over time, possibly breaking database wrappers used in mediator and link integration systems immediately. A local stable copy of a database can avoid breaking the system immediately whenever changes in source databases occur. Parsers used in a data warehouse can be adapted for changes more easily in this way.

Reliance on resources

Data warehouses do not need to rely on a working internet connection or the availability of resources at the service providers (remote server uptime and load), whereas link integration, mediator systems and Mashups are always as slow as their slowest data resource.

Handling of proprietary data

Data warehouses can easily be enriched with proprietary in-house data without the need for additional infrastructure (own web-service etc.). The data sources are downloaded once and integrated data is kept locally. This avoids that access and request pattern on public databases can be traced by the database providers. This is especially important in the pharmaceutical industry where access statistics to public information resources on the web could be used to reveal current pharmaceutical targets of interest, which could impose a significant risk to a pharmaceutical company.

Workflows

ONDEX as a data warehouse incorporates aspects of Workflows (see 5.1.5 "Workflows"). This is to facilitate both control of the data integration process and direct manipulation of the integration data structure used. The latter allows for a flexible approach, adding additional data resources available as web-services to ONDEX using a Workflow to transform data directly into ONDEX. The integration data structure can be seen as an explicit data model and storage for Workflows, which obliterates the need for on-the-fly construction of a data model otherwise common practise in Workflows.

7.3 ADDRESSING THE CHALLENGES

Six challenges were presented in Chapter 3 and are summarized in Table 1. This section briefly presents each challenge again. It shows how the challenge was addressed, what was achieved and highlights relevant aspects of the ONDEX system.

First challenge – representing biological data intuitively as a graph or network

Biological data can usually be seen as graphs or networks, in which the nodes represent biological entities (for example genes, proteins) and the edges represent their relationships (for example encodes, interacts with). To show that this approach is commonly accepted in the life sciences community a number of existing data integration and analysis frameworks have been presented in 2.3 "Survey of current data integration systems".

The ONDEX integration data structure (see 4.1 "ONDEX integration data structure") uses a graph-based approach to capture integrated biological data. It incorporates features of semantic networks and conceptual graphs. It was implemented using object-oriented programming techniques (see 5.2 "Implementing integration data structure"). It is able to handle graphs containing millions of entries. Aspects of transforming biological data into the integration data structure have been discussed (see 5.1 "System design"). A novel exchange format *OXL* for the exchange of integrated biological networks was introduced (see 4.3 "Exchanging integrated data").

Throughout all four use cases (see 6 "Use cases") data has been successfully integrated in ONDEX as a graph of concepts and relations. This graph has been visualised using the ONDEX Visualisation ToolKit (OVTK) [2]. Such visualisation helped our biologist colleagues and other users to better understand the biological knowledge contained within the graph. Relationships between biological entities (for example enzymes catalysing reactions) were intuitively understood using a graph-based representation.

7 Conclusion and outlook

Second challenge – overcoming the syntactic and semantic heterogeneities between data sources

Biological data is buried in hundreds of databases [141] and millions of publications. To access this vast amount of data one has to overcome syntactic and semantic challenges. Syntactic challenges include for example different data exchange formats or spelling mistakes in gene names. Semantic problems exist when same things are called differently, for example multiple gene names for one and the same gene; or different things are called the same, for example "ear" can be the human ear or a part of a wheat plant.

To address syntactic challenges between data sources, the ONDEX system has been designed to be easily extendible (see 5.1 "System design"). Adding new parsers for different data exchange formats is possible with minimum effort and was demonstrated in the last use case (see 6.4 "Analysis of social networks"). Semantic problems in the integrated data are addressed by mapping methods (see 4.2 "Data alignment"). These methods are based on attributes of concepts like accession numbers, synonyms or biological sequence information. A novel algorithm called *StructAlign* was introduced to overcome some of the semantic heterogeneities. To do so *StructAlign* considers the network neighbourhood of concepts in the graph. This algorithm together with two other mapping methods (accession based and synonym mapping) was evaluated in terms of precision, recall and F_1-score compared to a gold standard and between each other. It was shown that *StructAlign* can be particularly useful for the identification of equivalent concepts in the absence of accession numbers on concepts in biological data. The best results were obtained by combining all three presented mapping methods. The problem of not having reliable "gold standards" for evaluating mappings across biological data sets was discussed.

7 Conclusion and outlook

The first use case (see 6.1 "Improving genome annotations for *Arabidopsis thaliana*") integrated a variety of data sources including proteomics data from UniProt, protein structures from PDB, protein families from PFAM, pathways from AraCyc and classification information from Gene Ontology together in ONDEX. Similar and equivalent entities between these data sources were identified using mapping methods to address semantic heterogeneity in the integrated data. The result of this integration was a connected network which helped to improve genome annotations for *Arabidopsis thaliana*. The third use case (see 6.3 "Constructing a consensus metabolic network for *Arabidopsis thaliana*") builds on the evaluation results of the mapping methods presented. It shows that by overcoming representational differences in data sources (here between KEGG and AraCyc) using mapping methods it was possible to construct a consensus network. Such network provides a more comprehensive view on biological knowledge than each data source alone.

Third challenge – provide a semantical consistent view on integrated information

Flexible knowledge representation is a prerequisite for handling different types of biological information, for example metabolic networks, protein interaction networks or gene regulatory networks.

The metadata describing semantics or meaning of concepts and relations in the ONDEX integration data structure (see 4.1 "ONDEX integration data structure") can be defined by the user of the system. To support the user, the Metadata Editor (see 5.1.1 "Knowledge modelling and domain independence") was created. A visualisation of metadata in ONDEX is called a meta-graph. A meta-graph displays the concept classes and relation types used in the current integrated data, much like an Entity-Relationship (ER) diagram does for relational databases. It provides the user with a familiar overview of what is contained in the integrated data without requiring the user to inspect the actual instances in the graph.

7 Conclusion and outlook

In ONDEX a domain model is a set of metadata for a particular domain of knowledge. A domain model providing a semantical consistent view on information for biological knowledge of pathway data was suggested (see 5.1.2 "Formulating a consensus domain model in biology"). This domain model was successfully used and extended as part of the first use case (see 6.1 "Improving genome annotations for *Arabidopsis thaliana*"). The third use case (see 6.3 "Constructing a consensus metabolic network for *Arabidopsis thaliana*") showed how mapping methods, filtering and transforming can be used to change the view on integrated data. Starting from a pathway centric view of the data sources (here KEGG and AraCyc), it was transformed into a consensus metabolic network view of only reactions and metabolites as concepts in the graph.

Fourth challenge – keep track of provenance during integration process

Provenance about the data and integration steps performed is required to build trust in the data integration process and to enable further analysis of the data integration results.

Provenance in ONDEX is captured in multiple ways. Concepts in the ONDEX integration data structure are marked with their originating data source (see 4.1 "ONDEX integration data structure"). Both concepts and relations have types of evidence assigned. Evidence types reflect how entities in the integrated data were established. During the integration process links between similar or equivalent entries of different data sources are made explicit using mapping methods (see 4.2 "Data alignment"). Mapping methods assign scores to relations created. These scores can provide a measure of confidence for inferred relations. The integration process in ONDEX is controlled via workflows (see 5.1.5 "Workflows"). Workflows provide provenance about what integration steps were performed. A complex workflow was presented as part of the first use case (see 6.1.4 "Data integration pipeline"). This complex workflow shows in detail how the integrated dataset was created. The other use cases use simpler workflows.

7 Conclusion and outlook

Fifth challenge – domain independent approach to data integration

Data integration remains an unresolved problem also in other application domains outside the biological knowledge domain. For example, data integration is required for the analysis of consumer data, financial information, patent mining or social networks to name a few. Methods and principles applicable in each of these separate application domains could prove valuable for another.

To show that the methods and principles presented in this thesis can also be applied to other application domains, a proof of concept for social networks (see 6.4 "Analysis of social networks") was performed. The flexibility of the ONDEX integration data structure (see 4.1 "ONDEX integration data structure") made it possible to define a new set of metadata for this new application domain in ONDEX. Existing data integration (for example accession based mapping) and analysis (for example betweenness centrality) methods could readily be used with social network data. Visualisation of the graph structure and analysis results for social networks helped to gain insights into the social community represented. This proof-of-concept showed how easy it was to apply data integration, analysis and visualisation techniques developed as part of the ONDEX system to other application domains.

Sixth challenge – create a robust, usable and maintainable framework for data integration

ONDEX has been designed as an open-source framework for data integration in the life sciences. As part of the ONDEX SABR project (see 6.5 "ONDEX SABR project and its applications") major software engineering efforts did take place to increase the robustness, usability and maintainability of ONDEX.

It is now implemented as a platform independent JAVA application (see 5.1 "System design") using only proven open-source software components (for example Apache Lucene and Oracle Berkeley JAVA DB). A clear defined JAVA API is provided for the ONDEX integration data structure (see 5.2 "Implementing integration data structure"). Plug-in interfaces for all ONDEX components like parser or mapping methods make it easy to extend the system with new functionality. The *OXL* exchange format has been designed to satisfy the requirements (see 4.3.2 "Requirements for exchanging integrated data sets") derived from an analysis of existing exchange formats. Tool support is readily available for *OXL* and transformations into other representations like HTML or SQL have been feasible.

One database available as *OXL* is PHI-base. PHI-base was used as part of a comparative genomics analysis successfully predicting possible pathogenicity genes in newly sequenced organism using ONDEX (see 6.2 "Prediction of potential pathogenicity genes in *Fusarium graminearum*"). Other export formats available in ONDEX include for example SBML as shown in the third use case (see 6.3 "Constructing a consensus metabolic network for *Arabidopsis thaliana*"). This enables easy exchange of data between ONDEX and other Systems Biology tools.

7.4 COMPARISON WITH RELATED WORK

The methods and principles in this thesis have been implemented in the ONDEX data integration framework. BNDB with BN++ (see 2.3.3) is the most similar system to ONDEX according to data representation and system architecture. BNDB and ONDEX represent biological data as networks using a data warehouse approach. Both systems provide a server backend for data integration tasks and a front-end for visualisation, exploration and analysis of integrated data. However the systems differ in the conceptual approach to how data is modelled. BNDB uses the three distinct node features Event, Role and Participant to characterise biological entities. Extending these predefined semantics of BNDB is only possible by programmatically introducing new classes as subclasses of the core data structure. On the other hand ONDEX uses an approach similar to semantic network (see 8.2.1 "Semantic network") combined with a flexible type system on node and edges (see 4.1 "ONDEX integration data structure"). In ONDEX a structured XML file is used to initialise the semantics (metadata) of nodes and edges on the graph. This facilitates easy adaptation of the ONDEX system for use in a different knowledge domain by changing the semantics associated with the graph.

BNDB and ONDEX also differ with respect to how data is integrated. BNDB uses horizontal data integration where similar data from different sources is merged together during the import step. ONDEX, however, provides better means to track provenance of integrated data at each single step, i.e. data from different sources is first imported as different entities into the graph and only as a second step mapping methods (see 4.2 "Data alignment") are used to identify and later on merge similar biological concepts. This approach provides the user with more flexibility and choice to how data will be identified as equivalent and an easy way to extend and introduce new mapping methods. Additionally, it is possible to intercept the integration workflow (see 5.1.5 "Workflows") in ONDEX at each step and manipulate the integration data structure directly using its object oriented JAVA API.

SEMEDA (see 2.2.1) uses a federated approach to data integration and enables the user to search across the integrated resources on a webpage. SEMEDA does not have a front-end application which provides graph analysis and visualisation capabilities similar to the OVTK (ONDEX Visualisation ToolKit). ONDEX and SEMEDA do share some basic idea of how ontologies help to support data integration. In SEMEDA, ontologies are used to assign semantics to database attributes. In ONDEX, metadata on concepts and relations can be an ontology describing what classes of concepts and types of relations exist in the graph. The technical implementations between the two systems differ significantly. SEMEDA is implemented as a database schema on top of a relational database. The ONDEX integration data structure is implemented as an object oriented JAVA API. This API is independent of its actual persistent implementation, i.e. choice of database or storage system.

Other tools like Biozon (see 2.3.2) and STRING (see 2.3.4) focuses on providing readily integrated biological networks to the user via a web interface. BNDB / BN++, Proton (see 2.2.2) and ONDEX enable a computer literate user with means to assemble integrated datasets on his / her own. Visual Knowledge / BioCAD (see 2.3.1) or NeAT (see 2.3.5) emphasise biological pathways and networks analysis combined with a more ad-hoc approach to loading data from different sources. The ONDEX system provides access to integrated data in ONDEX via SOAP based web services and a powerful front-end application OVTK, which shows similarities to BN++.

The ONDEX workflow API as well as the integration data structure API have been exposed to the Taverna workbench (http://www.mygrid.org.uk) via web services which allows for seamless integration of the ONDEX data model with other existing web service resources, for example text mining tools, and the reuse of existing workflow based resources for data integration from sources like MyExperiment (http://www.myexperiment.org). The ONDEX framework can therefore be seen as a collection of seamlessly integrated tools for data integration which can be flexible

7 Conclusion and outlook

combined to address the challenge of data integration in the life sciences.

7.5 OUTLOOK

Although the ONDEX data integration framework is being successfully used in Systems Biology centres throughout the United Kingdom, it still requires a trained bioinformatician to apprehend the complexity of the system and adapt it to the relevant application case. Making a data integration and analysis system available on the web could improve uptake and ease of use as shown for example with "NeAT" (see 2.3.5). Therefore current efforts are being based on Web 2.0 technology to improve the accessibility and usability of the ONDEX system. This would not only enable bioinformatician but also computer literate biologists to easily compose workflows to suite their integration needs. This new platform would be based on the Google Gadget API (http://code.google.com/apis/gadgets/). Google Gadgets have been successfully used in creating network applications for the social analysis domain. The webpage to be created would present main ONDEX functionality wrapped into Google Gadgets which could be used as building blocks for analysis workflows. Predefined workflows would be readily available for reuse and would provide examples for all aspects of possible analysis. The analysis would be helped by a variety of pre-canned biological data sources in ONDEX format exposed via webservices. The system would support different levels of user authentication. It would also be possible to install it on a dedicated server for integration of proprietary and publicly available data.

Furthermore, with emerging new technologies and new data formats in the future it will become necessary to adapt the ONDEX data integration framework accordingly. The integration data structure (see 4.1 "ONDEX integration data structure") has been designed to cope with such changes in the future. Aggregating more and more aspects of biological knowledge into an integrated view could leverage our understanding of biology by providing a more and more

7 Conclusion and outlook

complete picture across all the different disciplines of biological research (for example combining genomics with proteomics, metabolomics and transcriptomics).

Although the data warehouse approach used by ONDEX satisfies our current data integration needs, the high costs involved in keeping the data warehouse up-to-date has opened up discussions involving other approaches. A hybrid of a data warehouse and mediator / federated system could help to address the problems of keeping data up-to-date. Such an approach would query the data sources on regular intervals automatically to feed the new data into the data warehouse. The data warehouse would still be used to process the user queries to keep the advantage of being independent of resource bottlenecks at the data providers. This can be seen as a sophisticated proxy system around ONDEX with features of a mediator / federated data integration system.

With maturing of standards and tools for semantic web, there has been an increasing need to take such technologies into account for ONDEX too. For example more and more data source providers start to supply their data in RDF using unification cross-references to other data sources. Such unification cross-references state exactly whether two information entities are the same in different data sources. Considering this information in a consistent way could ease the burden of identifying equivalent entities between data sources. RDF databases, so called triple stores, provide means to integrate different RDF documents together based on unification cross-references. Directly utilizing such RDF triple store for ONDEX would allow direct unification to happen automatically across different RDF documents. Nevertheless the semantic differences between data sources would still persist and will require transformation steps from the original representation into the domain model used by ONDEX. Using triple stores and RDF cannot solve all the challenges that data integration has to face. It can leverage the entry point at which methods for resolving syntactic and semantic conflicts between data sources have to work, but does not obliterate the need for them completely.

Closing remarks

"A valuable measure of success will be the delivery of scientific results that can be attributed to the use of the ONDEX framework." [Personal communication with Prof. C.J. Rawlings, Rothamsted Research, UK].

The ONDEX SABR project so far has been a milestone on the way to achieving this success. Within the UK life sciences community ONDEX has gained more recognition and generated a lot of interest. This echoes in the numerous requests and collaborations we have had since the start of the project. Although immediate biological discoveries are difficult to correlate directly with ONDEX, we are now in a position where data integration and analysis performed using the ONDEX data integration framework contributed noticeable to the process of biological discovery.

7 Conclusion and outlook

8 GLOSSARY

8.1 DEFINITIONS

8.1.1 GRAPH THEORY

The following definitions were extracted from [149] and are presented here for completeness and to assist the reader to understand this thesis.

GRAPH AND NETWORK: A graph is a symbolic representation of a network and of its connectivity. It implies an abstraction of reality so it can be simplified as a set of linked nodes. A graph G is a set of vertex (nodes) v connected by edges (links) e. Thus G = (v, e).

PATH: A sequence of links that are travelled in the same direction. For a path to exist between two nodes, it must be possible to travel an uninterrupted sequence of links.

CHAIN: A sequence of links having a connection in common with the other. Direction does not matter.

CYCLE: Refers to a chain where the initial and terminal node is the same and that does not use the same link more than once.

ROOT: A node *r* where every other node is the extremity of a path coming from *r* is a root. Direction is important.

TREE: A connected graph without a cycle is a tree. A tree has the same number of links as nodes plus one. ($e = v\text{-}1$). If a link is removed, the graph ceases to be connected. If a new link between two nodes is provided, a cycle is created. A branch of root *r* is a tree where no links are connecting any node more than once.

FOREST: A forest is a disjoint union of trees.

8.1.2 Properties on Relations

The following definitions for properties of relations were taken from [150].

REFLEXIVE: A relation R on a set X is reflexive if $(x,x) \in R$ for every $x \in X$.

SYMMETRIC: A relation R on a set X is symmetric if for all $x, y \in X$, if $(x,y) \in R$, then $(y,x) \in R$.

ANTI-SYMMETRIC: A relation R on a set X is anti-symmetric if for all $x, y \in X$, if $(x,y) \in R$ and $(y,x) \in R$, then $x = y$.

TRANSITIVE: A relation R on a set X is transitive if for all $x, y, z \in X$, if (x,y) and $(y,z) \in R$, then $(x,z) \in R$.

8.1.3 Object-oriented Development and UML

The method of Object-oriented analysis and design (OOAD) can be used to model information systems. It has been introduced by Booch [151]. Booch describes the process of object-oriented development with macro and micro processes.

The macro processes are:
- Establishment of core requirements (conceptualisation)
- Development of a model of desired behaviour (analysis)
- Creation of an architecture (design)
- Evolution of the implementation (evolution)
- Management of the post-delivery evolution (maintenance)

The micro processes are:
- Identification of classes and objects at a given level of abstraction
- Identification of the semantics of these classes and objects
- Identification of the relationships among these classes and objects
- Specification of the interface and then implementation of these classes and objects

The development process does not consist of exactly specified steps nor has a fixed order. It is sufficiently well-defined as to offer a predictable and repeatable process for the mature software development [151]. Adaptation of this process to the specific software project is necessary.

The *de facto* standard notation to describe object-oriented software is the Unified Modelling Language (UML) [152]. UML is used to illustrate developed solutions, for example by using UML class diagram. One initiator of UML is the already mentioned Gary Booch, another is James Rumbaugh, who is also the founder of an object-oriented software development method (Object-Modelling Technique (OMT) [153]). UML is the result of merging notations of both methods. Since 1994 UML has been continuously extended and improved, it is now available in Version 2.0.

8.1.4 ASPECT-ORIENTED SOFTWARE DEVELOPMENT

The following definition for Aspect-oriented software development (AOSD) is given in [154] and has been taken from [40]. "In computing, Aspect-oriented software development (AOSD) is an emerging software development technology that seeks new modularizations of software systems... AOSD allows multiple concerns to be expressed separately and automatically unified into working systems.

Traditional software development has focused on decomposing systems into units of primary functionality, while recognizing that there are other issues of concern that do not fit well into the primary decomposition. The traditional development process leaves it to the programmers to code modules corresponding to the primary functionality and to make sure that all other issues of concern are addressed in the code wherever appropriate. Programmers need to keep in mind all the things that need to be done, how to deal with each issue, the problems associated with the possible interactions, and the execution of the right behaviour at the right time. These concerns span multiple primary functional units within the application, and often result in serious problems faced during application development and maintenance. The distribution of the code for realizing a concern becomes especially critical as the requirements for that concern evolve — a system maintainer must find and correctly update a variety of situations.

Aspect-Oriented Software Development focuses on the identification, specification and representation of crosscutting concerns and their modularization into separate functional units as well as their automated composition into a working system."

8.2 KNOWLEDGE REPRESENTATION

Knowledge Representation (KR) is one of the central elements of Artificial Intelligence (AI), and contributes to all sorts of problem solving [155]. Sowa [156] defines it as the application of logic and ontology to the task of constructing computable models for some domain. He also states that AI design techniques have converged with techniques from other fields, especially database and object-oriented systems. He mentions rules, frames, semantic networks, object-oriented languages, Prolog, Java, SQL, Petri nets and the Knowledge Interchange Format (KIF) as major knowledge representations. Additionally, conceptual graphs are an important notation in knowledge representation.

Furthermore, knowledge representation can be described as organised by the five principles of Davis, Schrobe, and Szolovits [157]: a knowledge representation is: (1) "a surrogate"; (2) "a set of ontological commitments"; (3) "a fragmentary theory of intelligent reasoning"; (4) "a medium for efficient computation"; (5) "a medium of human expression".

8.2.1 SEMANTIC NETWORK

A semantic network represents information or knowledge by nodes and edges in a graph, which can be directed or undirected. Here, a node represents a concept and an edge represents a relationship [45]. Semantic networks are often used as a form of knowledge representation.

In a semantic network the meanings of words are defined through being embedded in a connected network of other meanings (see Figure 71 from [158]). Knowledge is validated and acquires meaning through correlation with other knowledge [159]. The nature of the connections within a semantic network can be described as being associative, as well as qualitative and purposeful. Therefore, the connections or links within a semantic network have semantic value.

8 Glossary

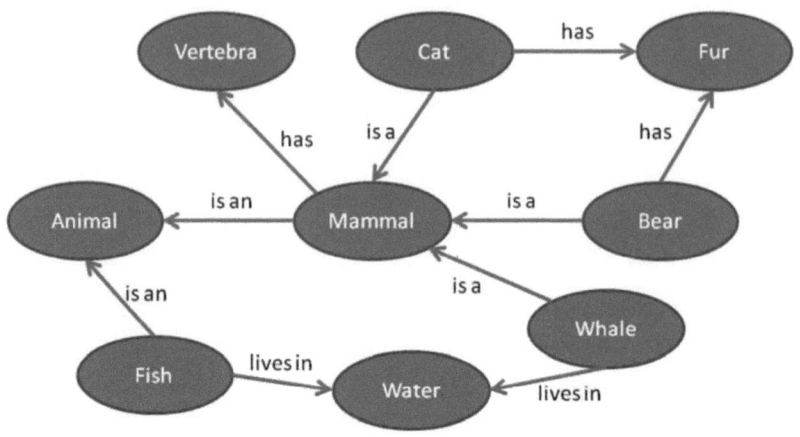

FIGURE 71 EXAMPLE OF A SEMANTIC NETWORK DEPICTING CLASSIFICATION OF DIFFERENT MAMMALS

8.2.2 CONCEPTUAL GRAPHS

Conceptual graphs are a two-dimensional form of logic that is based on the semantic networks of AI and the logical graphs of C. S. Peirce. A conceptual graph g is a bipartite graph that has two kinds of nodes called concepts and conceptual relations respectively. A simple example (taken from [156]) of a conceptual graph in linear notation is: [Cat] -> (On) -> [Mat]. [Cat] and [Mat] are concepts and (On) is a conceptual relation. The concept [Cat] by itself simply means 'There is a cat'.

Conceptual relations are directed. An inverse of a conceptual relation can exist, which can be used to infer the opposite direction of the relation when the forward link is made. For the above example the inverse of the conceptual relation (On) would be (Below) allowing inferring the conceptual graph: [Mat] -> (Below) -> [Cat].

"Every concept has a *type t* and a *referent r* ... In the concept [Bus], 'Bus' is the type and the referent is blank, ... In the concept [Person: John], 'Person' is the type, and the referent 'John' is the name of some person." [156]

Sowa [156] states that by representing logic in a form that is close to natural language, conceptual graphs can serve as an intermediate language for mapping to lower-level languages like SQL. As complete representations for logic, conceptual graphs and predicate calculus are general enough to represent everything that can be represented in a Petri net, a timing diagram, or any other notation for discrete processes.

8.2.3 ONTOLOGY

"Ontology ... is the study of existence, of all the kinds of entities—abstract and concrete—that make up the world ... The two sources of ontological categories are observation and reasoning ... A choice of ontological categories is the first step in designing a database, a knowledge base, or an object-oriented system." [156]

All forms of human knowledge can be encoded by Ontology in such a way that the knowledge can be used. Object orientation and generalisation help to make the represented knowledge more understandable to humans, reasoning and classification help make a system behave as if it knows what is represented [160]. Ontology consists of axioms (concepts and relationships between concepts) that define things. In knowledge representation, ontology has become the defining term for the part of a domain model that excludes the instances, yet describes what they can be. Ontological analysis is the process of defining this part of the model.

8.2.4 CONTEXTS

Contexts support the representation of encapsulated objects of object-oriented systems. In conceptual graphs, contexts are represented by concept boxes that contain nested graphs that describe the referent of the concept, i.e. a context is defined as a concept whose referent field contains nested conceptual graphs [156]. Several levels of nested conceptual graphs may exist. A context can be extended, in which case the nested graph of the referent is shown.

8 Glossary

8.2.5 DOMAIN KNOWLEDGE

Most generally, domain knowledge is the knowledge which is valid and directly used to refer to an area of human endeavour, an autonomous computer activity or other specialized discipline. Domain knowledge is used and developed by specialists and experts in their own domain. If the terms *domain knowledge* or *domain expert* are used, emphasis is given to a specific domain which is part of the discourse, interest or problem [161].

8.2.6 DOMAIN MODELLING

Figure 72 (from [160]) shows a framework for discussing domain modelling according to [162]. The amorphous shape labelled *Domain Knowledge* refers to the knowledge possessed by the domain expert that must be encoded in some fashion. This knowledge is not well defined and is fairly difficult for others to access. The box labelled *Meta Model* refers to the KR formalism, typically a KR language, that will be used as the *symbol level* [155] for the machine representation of this knowledge. The box labelled *Instantiation* refers to the *process* of taking the domain knowledge and physically representing it using the meta-model, this process is sometimes referred to as *knowledge acquisition* [163]. The box labelled *Domain Model* refers to the knowledge-base which results from the instantiation.

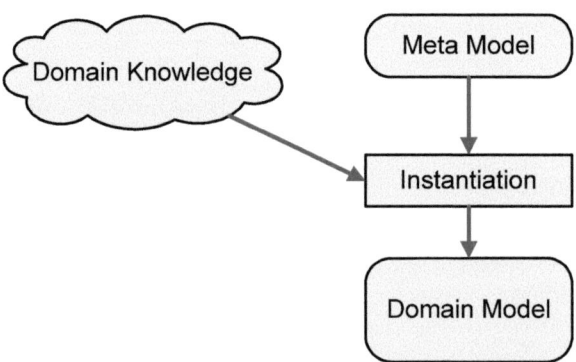

FIGURE 72 FRAMEWORK FOR DISCUSSING DOMAIN MODELLING

The *Meta Model* is independent of the *Domain Knowledge*. It can be instantiated with different *Domain Knowledge* resulting in different *Domain Models*. Therefore the *Meta Model* can be called "domain independent". The degree of domain independence of the *Meta Model* can vary according to the KR formalism used, i.e. some KR languages (for example Petri nets) may already have a tendency or being predisposed towards the use in a particular domains of knowledge, whereas others may be of a very general character (for example object-oriented programming or SQL).

8.2.7 HIERARCHY AND TAXONOMIES

Knowledge in ontology can be structured as a hierarchy. The simplest yet most common form of a hierarchy for ontology is a taxonomy. Taxonomy is a collection of Controlled Vocabulary terms organised in a hierarchical structure. Relationships in taxonomy are specified in terms of generalisation and specialisation, i.e. each term in taxonomy is in one or more parent / child (broader / narrower) relationships to other terms in the taxonomy.

Concept taxonomies specify what types of concepts can exist in a domain model. Relationship taxonomies specify the generalisation / specialisation hierarchy between conceptual relations. For example (taken from [160]), ontology for describing cars might include conceptual relations called *has-engine*, *has-seats*, and *has-headlights*, which relate concepts that represent cars to concepts that represent engines, seats, and headlights, respectively. The conceptual relation *has-parts*, then, could be expressed as the generalisation of all these conceptual relations, and the result is that all the values of all the more specialised conceptual relations would also be values of the more general conceptual relation.

8.2.8 CONTROLLED VOCABULARY

According to [147] controlled vocabulary can be defined as a regularised or standardised list of terms representing words drawn from natural language used to increase uniformity in indexing or information retrieval, and to enhance the sharing of information among institutions, individuals, and within and across disciplines. It includes subject headings, thesauri, and Taxonomies.

8.3 TERMS USED IN ONDEX

Alignment method – see mapping

API – Application programming interface. API is a collection of classes and methods exposed publicly in a software system. The API hides the actual implementation of the classes and methods.

Concept – a node in the graph-based ONDEX data integration structure representing a general idea about a subject of discourse. For example a concept for a certain protein would represent all possible instances of that kind of protein which could exist in the reality.

Concept accession – can be assigned to a concept and represents a cross reference for this concept in another data source. A concept accession usually is a triple (accession, data source, ambiguous) with ambiguous true if there is a conceptual difference between the current concept and the concept in another data source. For example, protein concept *P00330* would have ("2HCY", "PDB", false) and ("S000005446", "SGD", true).

Concept class – an identifier representing the class of a concept. Concept classes form a taxonomy and exactly one concept class is assigned to one concept. For example the protein concept *P00330* would have the concept class "Protein".

Concept name – can be assigned to a concept and usually captures human readable synonyms for a concept. A concept name is a tuple (name, preferred) with preferred true if this name is an exact or preferred synonym for the concept. For example, protein concept *P00330* would have ("ADH1_YEAST", true) and ("ADH1", false).

Context – concepts and relations may have other concepts assigned to them to define some sort of explicit grouping or structuring. For example, a reaction and an enzyme concept could have a pathway concept assigned as context to show that both enzyme and reaction belong to the same biological pathway.

8 Glossary

CV (Controlled Vocabulary) – an identifier for a particular data source used for a concept when identifying the data source this concept has been extracted from and on a concept accession to identify which data source the accession is pointing to. For example, concept *P00330* would itself have CV "UniProt" and the concept accession ("2HCY", "PDB", false) contains CV "PDB".

Evidence type – an identifier to describe provenance about how something has been established assigned as lists to both concepts and relations. For example, concept *P00330* could have evidence type "IMPD" assigned as it has been simply "IMPorted from Database".

Export – is a part of the system which exports some or all contents of the integration data structure into another computer readable format, usually at the end of a workflow containing its results. For example at the end of a workflow the resulting data set can be exported to OXL for loading it into the OVTK.

Filter – extracts sub-graphs from the integration data structure defined by properties associated with concepts and relations. For example, a filter extracts only sub-graphs containing concepts with concept classes "Reaction" and "Metabolite" from a more complex representation of a biological pathway made of many more concept classes.

GUI – (graphical user interface) enables the user to interact with a system

Importer – see Parser

Mapping – mapping methods create new relations between two concepts based on some similarity measure of properties of the two concepts. In the simplest case a mapping method could check if two concepts have bi-directional concept accessions, for example concept *P00330* from UniProt and concept *2HCY* from PDB would be identified to represent the same protein concept and a new relation of type "equ" (equivalent) would be created between them.

ONDEX – the term ONDEX was introduced by Jacob Köhler and Andre Skusa in 2004 as an acronym for "ONtological text inDEXing", however it has lost this meaning over the years and is now no longer an acronym. Also has the ONDEX system moved away from being a text-mining centred tool towards a more general framework for data integration in the life sciences. Therefore ONDEX has become a "trade name" for the data integration framework presented in this thesis.

OVTK (ONDEX Visualisation ToolKit) – is the front-end application mainly used in the ONDEX data integration framework which allows for the ONDEX data integration structure to be visualised a graph which is highly customisable and interactive.

OXL – the "Ondex eXchange Language", a XML based format for the exchange of integrated data sets between components of ONDEX and other data sources.

Parser – a parser takes a data source and extracts certain aspects of information. In ONDEX a parser does the translation of the information in the data source into the ONDEX data integration structure, here a data source could be flat-files (for example OXL), a whole database (for example KEGG) or even web-services. This translation of information is usually not "lossless".

Relation – an edge in the graph-based ONDEX data integration structure representing a general idea about the relationship between two concepts. For example two protein concepts could have an interaction relationship between them represented by a relation.

Relation type – is an identifier representing the type of a relation. Relation types are forming taxonomy and exactly one relation type class is assigned to one relation. For example the relation type on a relation describing the interaction between two proteins could be "it_wi" ("interacts with").

Transformer – as part of the ONDEX system is used to perform arbitrary manipulation to concepts and relations in the integration data structure, which will transform their representation away from the state generated by a parser. For example, a transformer could merge or collapse two concepts previously identified as being equivalent by a mapping method into one single concept and thus remove redundancy from the integration data structure.

Workflow – a workflow describes how the different components in a system should work together and in which order to achieve a particular task. In ONDEX the components of the system include for example parser, mappings, transformer, filter or exports. A workflow could be for example: start "parser A", next "parser B", next "mapping between A and B", next "export A and B", end. ONDEX supports a XML based description for these workflows and a GUI.

9 APPENDIX

9.1 LIST OF DATA FORMATS SUPPORTED BY ONDEX

9.1.1 IMPORT

Data import in ONDEX is facilitated through use of different parsers. A distinction is made between generic format parsers and database / tool specific parsers.

Generic format parsers include:
- FASTA (http://www.ncbi.nlm.nih.gov/blast/fasta.shtml)
- GO (OBO 1.2 format, http://www.geneontology.org/GO.format.shtml)
- PSI-MI (version 2.5, http://www.psidev.info/index.php?q=node/60)
- SBML (version 2, http://sbml.org/Main_Page)
- Tab-delimited

Database/tool-specific Parsers:
- AraCyc (http://www.arabidopsis.org/biocyc/index.jsp)
- AtRegNet (http://arabidopsis.med.ohio-state.edu/moreNetwork.html)
- BioCyc (http://biocyc.org/)
- BioGRID (http://www.thebiogrid.org/)
- Brenda (http://www.brenda-enzymes.org/)
- Cytoscape (http://www.cytoscape.org/)
- EcoCyc (http://ecocyc.org/)
- GOA (http://www.ebi.ac.uk/GOA/)
- Gramene (http://www.gramene.org/)
- Grassius (http://grassius.org/)
- KEGG (http://www.genome.jp/kegg/)
- Medline (http://medline.cos.com/)
- MetaCyc (http://metacyc.org/)
- O-GlycBase (http://www.cbs.dtu.dk/databases/OGLYCBASE/)

9 Appendix

- OMIM (http://www.ncbi.nlm.nih.gov/omim/)
- PDB (http://www.rcsb.org/pdb/home/home.do)
- Pfam (http://pfam.sanger.ac.uk/)
- Prolog (limited functionality / syntax)
- SGD (http://www.yeastgenome.org/)
- TAIR (http://www.arabidopsis.org/)
- TIGR (http://www.tigr.org/tdb/e2k1/ath1/)
- Transfac (http://www.gene-regulation.com/pub/databases.html)
- Transpath (http://www.gene-regulation.com/pub/databases.html)
- UniProt (http://www.uniprot.org/)
- WordNet (http://wordnet.princeton.edu/)

9.1.2 EXPORT

The ONDEX system supports two kinds of export, exporting the graph or network in a computer readable format or exporting the graphical representation (the image) of the network.

Network Data Export:
- DOT (http://www.graphviz.org/doc/info/lang.html)
- GraphML (http://graphml.graphdrawing.org/)
- Prolog
- SBML (limit functionality, http://sbml.org/Main_Page)
- Tab-delimited

Graphics Export:
- BMP
- EPS
- JPEG
- PDF
- PNG
- SVG

9.2 DATA INTEGRATION METHODS IN ONDEX

The following is a list of supported data integration methods in ONDEX with a short description each.

Accession based mapping: See 4.2.2.2 "Data integration methods and algorithms". Equivalent concepts in the graph are identified using concept accession data. If two concepts share the same unambiguous accession of the same data source, then a new relation with type "equivalent" is created between them. A score describing the proportion of matching accessions is assigned as an attribute on the relation.

BLAST based mapping: Performs an all-against-all BLAST [79] of sequence data associated with concepts in the graph. This mapping method creates new relations between concepts with type "has similar sequence" and the BLAST bit score assigned as an attribute on the relation.

Cross species sequence mapping: Uses the hardware accelerated sequence alignment platform TimeLogic Decypher (http://www.timelogic.com) to derive sequence similarity between concepts belonging to different organisms. This mapping can be seen as a specialisation of the BLAST based mapping.

EC2GO mapping: This mapping is a specialisation of the External2GO mapping and parses the ec2go mapping file (http://www.geneontology.org/external2go/ec2go) to create "equivalent" relationships between concepts of the Enzyme Nomenclature (EC)[69] and the Gene Ontology (GO)[71].

External2go mapping: See 4.2.2.3 "Other data integration methods". Loads mapping lists of concept identifiers to GO terms and creates relations between concepts of the Gene Ontology (GO)[71] and concepts of other data sources. Three different relation types will be created: "has function" to concepts of class "Molecular Function" from GO, "has process" to concepts of class "Biological Process" from GO and "is located in" to concepts of class "Cellular Location" from GO.

9 Appendix

GDS equality mapping: Creates an "equivalent" relation between concepts if they have particular equivalent GDS attributes, which can be specified. For example, such an attribute could be a CRC checksum of sequence data on a concept. Then if the sequence CRC checksum is the same between two concepts, it implies that the sequence of two concepts is the same and an "equivalent" relation can be created.

GO2GOSLIM mapping: An equivalent of the map2slim algorithm adopted for ONDEX, see http://search.cpan.org/~cmungall/go-perl/scripts/map2slim. The annotation is collected from child terms, which is adapted for subsumed children with S1-S2 where S1 and S2 are sets of GO children for GOSLIM (http://www.geneontology.org/GO.slims.shtml) terms t1 and t2 respectively and t2 is a child of t1.

Graph pattern mapping: Creates mappings based on relationships between end entries of a graph query pattern, i.e. pattern = Enzyme, Protein, Gene will map entries of concept class Enzyme only if a specified relationship exists between entries of concept class Gene reachable from these entries of concept class Enzyme. If no particular relationship is specified, entries are expected to be equivalent.

InParanoid mapping: Implements the InParanoid [139] algorithms for the identification of groups of orthologous and paralogous proteins in the graph. Confidence values are assigned to the new relations of type "ortho" and "para" created by this mapping method.

PFAM2GO mapping: This mapping is a specialisation of the External2GO mapping and parses the pfam2go mapping file (http://www.geneontology.org/external2go/pfam2go) to create "equivalent" relationships between concepts of the Protein family database (PFAM)[78] and the Gene Ontology (GO)[71].

Name based mapping: See 4.2.2.2 "Data integration methods and algorithms". Equivalent concepts in the graph are identified using concept names. Concept names are synonyms of concepts. If a certain threshold of concept names between two different concepts matches, then a new relation with type "equivalent" is created between them. A score describing the proportion of matching concept names is assigned as an attribute on the relation.

Sequence2pfam mapping: See 4.2.2.3 "Other data integration methods". Assigns protein domain functional information to protein sequences by first exporting the sequence data into a FASTA [77] file and then matching it against consensus sequences from a local PFAM database [78] using BLAST [79] or HMMER [80]. The created relations will have the bit score of the respective matches assigned.

StructAlign Mapping: See 4.2.2.2 "Data integration methods and algorithms". The StructAlign mapping algorithm considers the graph neighbourhood of concepts. A breadth-first search of a given depth (≥1) looks at matching concept names in the neighbourhood of the concepts under consideration. If at any depth one or more pairs of concepts which share synonyms are found, *StructAlign* creates a new relation between the concepts under consideration and assigns a score describing the search depth and proportion of matching concept names.

Text mining based mapping: Creates relationships between the concepts of a specified concept class and publications in ONDEX by searching for the occurrence of concept names in publication abstracts indexed by the Lucene (http://lucene.apache.org) full-text indexing system. The created relations have certain information retrieval scores, like TF-IDF (term frequency – inverse document frequency) assigned.

Transitive mapping: See 4.2.2.3 "Other data integration methods". Transitive relationships between concepts are inferred from existing relations. For example, if concept A is identified to be equivalent to concept B and concept B is known to be equivalent to concept C, then a new equivalent relationship between concept A and concept C is created by this mapping method.

9 Appendix

TABLE OF FIGURES

FIGURE 1 SYSTEMS BIOLOGY CYCLE OF EXPERIMENT, ANALYSIS, INSIGHTS, MODEL AND HYPOTHESIS TOGETHER WITH REQUIREMENTS FOR LARGE DATA FOR ANALYSIS OF EXPERIMENTAL RESULTS AND MODEL DEVELOPMENT... 2

FIGURE 2 OVERVIEW OF ONDEX SYSTEM USING A THREE STEP APPROACH OF 1) DATA INPUT (LEFT), 2) DATA INTEGRATION (MIDDLE) AND 3) DATA ANALYSIS (RIGHT). HIGHLIGHTED PARTS WILL BE THE MAIN TOPICS OF THIS THESIS; ALL OTHER PARTS WILL ONLY BE BRIEFLY DESCRIBED........ 3

FIGURE 3 DATABASE ATTRIBUTES CAN BE DEFINED BY LINKING THEM TO ONTOLOGY CONCEPTS (THICK ARROWS), 'ENAME' AND 'ENZ' ARE DEFINED IN THIS EXAMPLE AS THE SAME CONCEPT 'ENZYME', I.E. THEY BOTH CONTAIN ENZYME NAMES. 'ORG' CONTAINS ONLY INVERTEBRATES, WHEREAS 'SPEC' CONTAINS VERTEBRATES. 13

FIGURE 4 BY MAPPING SYNONYMOUS CONCEPTS OF CONTROLLED VOCABULARIES, IT IS POSSIBLE TO RELATE DATABASE ENTRIES THAT USE DIFFERENT TERMS FOR THE SAME THINGS. 14

FIGURE 5 DATABASE ATTRIBUTES WHICH ARE DEFINED AS THE SAME CONCEPT AND SHARE THE SAME CONTROLLED VOCABULARY AS THEIR DOMAIN CAN BE USED FOR CROSS-REFERENCING BETWEEN DATABASE ATTRIBUTES. 14

FIGURE 6 SEMEDA'S DATABASE QUERY INTERFACE. (1) ALL CONCEPTS FOR WHICH DATABASE ATTRIBUTES EXIST ARE LISTED. EACH OF THE ROUND ICONS REPRESENTS A DATABASE TABLE THAT CONTAINS AN ATTRIBUTE FOR THE CONCEPT. (2) AFTER SELECTING ONE OF THE ICONS, AN APPROPRIATE FORM OF THE RESPECTIVE TABLE MAY BE USED TO QUERY THE DATABASE. (3) THE RESULT SET IS EXTENDED WITH ICONS THAT ARE USED TO CROSS-REFERENCE OTHER DATABASES. 15

FIGURE 7 PROTON SYSTEM OVERVIEW DEPICTING THE USE OF METADATA (MIDDLE LEFT) FOR THE MODELLING AND VISUALISATION PROCESS (MIDDLE RIGHT) AND THE STORAGE INTERFACES OF THE SYSTEM (BOTTOM)... 16

FIGURE 8 VISUALISATION OF INTEGRATED DATA IN PROTON SHOWING DIFFERENT VIEWS ABOUT METABOLIC NETWORKS. 17

FIGURE 9 EXAMPLE OF A SEMANTIC NETWORK. CHARACTERISTICS AND BEHAVIOURS OF A SEMANTIC AGENT (SA) ARE DEFINED BY ITS RELATIONSHIPS (RE) WITH OTHER AGENTS. 20

FIGURE 10 SCREENSHOT OF SEMANTIC NETWORK (SN) SIMULATOR. UPPER PART DEPICTS THE PATHWAY VIEW; LOWER PART REVEALS THE LOCATION OF REACTANTS. 20

FIGURE 11 FRONT-PAGE OF BIOZON. CENTRE: SHOWING DIFFERENT OPTIONS TO START A SEARCH. BOTTOM RIGHT: SOME STATISTICS ABOUT THE INTEGRATED DATASETS AND DATA SOURCES... 22

FIGURE 12 OVERVIEW ARCHITECTURE BNDB / BN++. THREE LAYERS FROM BOTTOM TO TOP: DATABASE LAYER WITH IMPORTERS (RIGHT) FOR DIFFERENT DATA SOURCES, A WEB SERVER PROVIDING A SOAP INTERFACE (MIDDLE) AND FRONT-END APPLICATION BINA (TOP) WITH PLUG-INS.. 23

Table of figures

FIGURE 13 VISUALISATION FRONT-END BINA SHOWING AN EXAMPLE METABOLIC NETWORK WITH EC NUMBERS ATTACHED TO EDGES AND METABOLITES REPRESENTED ON NODES. 23

FIGURE 14 EXAMPLE PROTEIN INTERACTION NETWORK VIEW IN STRING WITH ADDITIONAL INFORMATION ABOUT SELECTED PROTEIN SHOWING IN INSET. 24

FIGURE 15 NEA-TOOLS RESULTS EXAMPLE AFTER A SUCCESSFUL "COMPARE-GRAPHS" OPERATION. DIFFERENT OPTIONS TO PROCEED ARE GIVEN AT THE BOTTOM OF THE RESULTS PAGE. 26

FIGURE 16 EXAMPLES OF BIOLOGICAL KNOWLEDGE AS GRAPHS: A.) PROTEIN INTERACTIONS, B.) METABOLIC PATHWAYS, C.) BIOLOGICAL ONTOLOGIES 32

FIGURE 17 INTEGRATION DATA STRUCTURE (HIGHLIGHTED) IS THE CENTRAL COMPONENT OF THE ONDEX DATA INTEGRATION FRAMEWORK 42

FIGURE 18 A SUBSET OF UNIPROT REPRESENTED AS A GRAPH IN THE ONDEX INTEGRATION DATA STRUCTURE WITH A PROTEIN CONCEPT (CIRCLE) CONNECTED TO CONCEPTS FOR DISEASE (TRIANGLE VIA "INVOLVED_IN" RELATION), ENYZME CLASSIFICATION (EC) (PENTAGON VIA "CAT_C" RELATION) AND PUBLICATIONS (SQUARE VIA "PUBLISHEDIN" RELATION). 44

FIGURE 19 EXCERPT FROM THE SEQUENCE ONTOLOGY SHOWING HIERARCHY OF CLASSES ROOTED IN DOMAIN SPECIFYING NODE "SEQUENCE ONTOLOGY" 45

FIGURE 20 OBO RELATION ONTOLOGY SHOWING DIFFERENT RELATION TYPES AND THEIR HIERARCHY ROOTED IN THE MOST GENERAL TYPE "RELATIONSHIP" 48

FIGURE 21 DATA INTEGRATION IN ONDEX CONSISTS OF 3 STEPS. 1) IMPORT AND CONVERSION OF DATA SOURCES INTO THE INTEGRATION DATA STRUCTURE OF ONDEX (DATA INPUT LEFT), 2) LINKING OF EQUIVALENT OR RELATED ENTITIES OF THE DIFFERENT DATA SOURCES (DATA INTEGRATION MIDDLE), 3) KNOWLEDGE EXTRACTION IN THE GRAPH BASED ANALYSIS COMPONENT (DATA ANALYSIS RIGHT). 59

FIGURE 22 WORKED EXAMPLE FOR STRUCTALIGN. DIFFERENT SHADES ARE USED TO DISTINGUISH DATA SOURCES. NODE SHAPE REPRESENTS DIFFERENT CLASSES OF CONCEPTS, SQUARE FOR ENZYMES AND CIRCLE FOR METABOLITES. ROUND ARROWS SHOW MATCHING SYNONYMS, WHEREAS VERTICAL ARROWS REPRESENT EXISTING KNOWLEDGE FROM DATA SOURCES AND HORIZONTAL ARROWS ARE CREATED BY STRUCTALIGN. 67

FIGURE 23 A) CHLOROPHYLL A BIOSYNTHESIS I PATHWAY FROM ARACYC (BOTTOM NODES) WITH CORRESPONDING SUBSET FROM KEGG (TOP NODES). NODES IDENTIFIED TO BE EQUIVALENT IN DARKER SHADE. B) META-GRAPH PROVIDING AN OVERVIEW OF THE INTEGRATED DATA, NODE COLOUR AND SHAPE DISTINGUISHES CLASSES; EDGE COLOUR DISTINGUISHES DIFFERENT RELATION TYPES. 78

FIGURE 24 SHOWS THE DIFFERENT COMPONENTS OF ONDEX AND HOW ONDEX MAKES USE OF THE NEW EXCHANGE FORMAT (OXL HIGHLIGHTED) AS WELL AS OTHER STANDARD EXCHANGE FORMATS. INTEGRATION RUNS CONSISTS OF THREE STEPS: INPUT OF DATA FROM

Table of figures

DIFFERENT DATA SOURCES (INCLUDING OXL), THE INTEGRATION PROCESS IN THE ONDEX INTEGRATION DATA STRUCTURE (INITIALISATION DATA VIA OXL), AND DATA ANALYSIS USING DIFFERENT TOOLS AND INTERFACES (DATA EXCHANGE USING OXL). 82

FIGURE 25 OXL XML SCHEMA – ONDEXDATA ELEMENT IS THE STARTING ELEMENT FOR OXL, THIS ELEMENT HAS DECISIVE CHARACTER ABOUT THE CONTENT OF THE OXL FILE, WHETHER DESCRIBING META-DATA OR AN ONDEX GRAPH. 88

FIGURE 26 OXL XML SCHEMA – METADATA ELEMENT CONTAINS ALL CLASSES FOR META-DATA CURRENTLY PRESENT IN ONDEX, NAMELY CV, UNIT, ATTRIBUTENAME, EVIDENCETYPE, CONCEPTCLASS AND RELATIONTYPE. 90

FIGURE 27 OXL XML SCHEMA – CONCEPT ELEMENT DESCRIBING PROPERTIES AND META-DATA OF ONE CONCEPT 92

FIGURE 28 OXL XML SCHEMA – RELATION ELEMENT DESCRIBING PROPERTIES AND META-DATA OF ONE RELATION 92

FIGURE 29 OXL METADATA EXCERPT FOR PR_BY RELATION TYPE, WHICH IS A SPECIALISATION OF THE MOST GENERAL RELATION TYPE "R". 92

FIGURE 30 RESULT TABLE FOR PHI-BASE ENTRY PHI:441. TOP PART IS A QUERY INPUT FORM TO COMPOSE ADVANCED QUERIES ON PHI-BASE, WHEREAS THE BOTTOM PART DOES CONTAIN THE LIST OF RESULTS. 95

FIGURE 31 EXCERPT OF OXL EXPORTED FROM PHI-BASE, HERE ONLY SHOWING THE FIRST CONCEPT WITH PHI-BASE ID PHI:441. 96

FIGURE 32 SCREENSHOT OF OVTK, DATA LOADED FROM OXL SHOWING RELATIONSHIPS OF PHI:441. ACTUAL GRAPH OF RELATIONSHIPS DISPLAYED IN CENTRAL "VISUALIZATION" FRAME. TOP RIGHT "METAGRAPH VIEW": OVERVIEW OF METADATA (CONCEPT CLASSES LIKE GENE, DISEASE, AND RELATION TYPES BETWEEN MEMBERS OF THESE CONCEPT CLASSES LIKE PRECEDED_BY, INTERACTING_WITH). LOWER RIGHT: CONCEPT CLASS COLOURS AND SYMBOLS. 97

FIGURE 33 OVERVIEW OF A TYPICAL ONDEX PIPELINE CONSISTING OF THREE PARTS: 1) PARSING HETEROGENEOUS DATA SOURCES INTO THE OBJECT BASED DATA MODEL (INTEGRATION DATA STRUCTURE) OF ONDEX; 2) IDENTIFYING EQUIVALENT AND RELATED ENTRIES AND CREATING NEW RELATIONS BETWEEN THEM USING MAPPING METHODS; 3) ANALYSING THE INTEGRATED DATA USING CLIENT TOOLS, FOR EXAMPLE THE ONDEX VISUALISATION TOOL KIT (OVTK) 102

FIGURE 34 INSTANTIATION OF META MODEL WITH DOMAIN KNOWLEDGE LEADS TO DOMAIN MODEL 103

FIGURE 35 ONDEX META DATA EDITOR IMPLEMENTED BY JOCHEN WEILE. DIFFERENT ONDEX METADATA IS AVAILABLE ON SEPARATE TABS, WHEREAS THE HIERARCHY OF META-DATA IS DISPLAYED AS A TREE ON THE LEFT. 104

FIGURE 36 META-GRAPH ARACYC DATABASE CONTAINING CONCEPT CLASSES FOR PROTEINS (CIRCLE), REACTIONS (OCTAAGON) OR PATHWAYS (STAR) CONNECTED BY DIFFERENT RELATION TYPES DISTINGUISHED BY SHADES. 106

FIGURE 37 META-GRAPH KEGG DATABASE CONTAINING CONCEPT CLASSES FOR PROTEINS (CIRCLE), REACTIONS (OCTAGON) OR PATHWAYS (STAR)

Table of figures

CONNECTED BY DIFFERENT RELATION TYPES DISTINGUISHED BY SHADES. 107
FIGURE 38 META-GRAPH UNIPROT DATABASE WITH CONCEPT CLASS FOR PROTEINS (CIRCLE) CONNECTED TO CONCEPT CLASSES OF ADDITIONAL CHARACTER, LIKE PUBLICATIONS (SQUARE) AND EC NUMBER (PENTAGON) 107
FIGURE 39 META-GRAPH OF THE COMBINED CONSENSUS DOMAIN MODEL ACROSS ALL THREE DATABASES ARACYC, KEGG AND UNIPROT 108
FIGURE 40 PATHWAY MAP ATH00010 FROM KEGG WITH ORGANISM SPECIFIC ENZYMES HIGHLIGHTED AND METABOLITES REPRESENTED BY SMALL WHITE CIRCLES. 111
FIGURE 41 SIMPLE ONDEX WORKFLOW XML. FIRST STARTING WITH AN EMPTY IN-MEMORY GRAPH, FOLLOWED BY A KEGG PARSER RUN FOR SPECIES CODE "ACB" AND FINALLY EXPORT OF RESULTS TO OXL. 116
FIGURE 42 ONDEX WORKFLOW MANAGER ALLOWS SELECTING AVAILABLE WORKFLOW COMPONENTS FROM THE LEFT AND ADDING THEM TO THE CURRENT WORKFLOW ON THE RIGHT. PARAMETER FORMS CAN BE FILLED IN FOR EACH COMPONENT. 116
FIGURE 43 UML CLASS DIAGRAM OF JAVA API 120
FIGURE 44 THREE TIER ARCHITECTURE OF THE ONDEX GRAPH IMPLEMENTATION STARTING WITH THE MOST GENERAL API, CONTINUING TO ABSTRACT CLASSES IMPLEMENTING SHARED FUNCTIONS AND CONCRETE IMPLEMENTATIONS AS MEMORY OR BERKELEY DB IN THE LAST TIER 122
FIGURE 45 LUCENEQUERYBUILDER IMPLEMENTATION – EXCERPT OF EXPOSED METHOD SIGNATURES. 128
FIGURE 46 OVERVIEW OF THE ONDEX DATA INTEGRATION SYSTEM WITH COMPONENTS RELEVANT TO THIS APPLICATION CASE 131
FIGURE 47 ONDEX META-GRAPH FOR THE ARACYC DATABASE, DIFFERENT CONCEPT CLASSES LIKE PATHWAY (STAR) OR REACTION (OCTAGON) CONNECTED VIA DIFFERENT RELATION TYPES LIKE "MEMBER IS PART" (DARKER SHADE) 132
FIGURE 48 PIPELINE FOR ANNOTATING PROTEIN SEQUENCES WITH GO TERMS, PROTEIN FAMILY INFORMATION AND PDB STRUCTURES. 1-5: INTEGRATING ARACYC, GOA AND UNIPROTKB; 6-9: ADDING PFAM-FAMILY INFORMATION TO PROTEINS; 10-12: MAPPING STRUCTURAL INFORMATION; 13-16: MAPPING GO TERMS TO PROTEINS. 134
FIGURE 49 AFTER THE "COLLAPSE FILTER" WAS APPLIED THE PROTEINS (CENTRE) WERE MERGED INTO ONE CONCEPT. 135
FIGURE 50 THE DASHED RELATION IS CREATED USING THE TRANSITIVE MAPPING. 136
FIGURE 51 RESULTING REPRESENTATION OF AN EXTRACT OF THE "LEUCOPELARGONIDIN AND LEUCOCYANIDIN BIOSYNTHESIS". AN ARACYC PATHWAY IN OVTK WITH GO, PFAM AND PROTEIN STRUCTURE INFORMATION. 137
FIGURE 52 EVALUATING PFAM BASED GO MAPPINGS TO AN *ARABIDOPSIS* REFERENCE SET FROM PUBLICLY AVAILABLE GOA FILES. 138
FIGURE 53 PHI-BASE, A.) FRONT-PAGE WITH QUERY MASK AND B.) RESULT PAGE FOR QUERY 'CANDIDA A*'. 145

Table of figures

FIGURE 54 WORKFLOW IMPORTS DATA FROM PHI-BASE AND A FASTA FILE INTO THE INTEGRATION DATA STRUCTURE. THE ONDEX IMPLEMENTATION OF THE INPARANOID ALGORITHM BASED ON BLAST MAPPINGS OF THE GENOMIC DATA IS RUN. THE INTEGRATION RESULTS ARE FILTERED AND EXPORTED IN XML. 146

FIGURE 55 OVTK SHOWING THE RESULT OF THE INTEGRATION WORKFLOW APPLIED TO NOVEL GENES FROM *FUSARIUM GRAMINEARUM* (RED SQUARES) AND ENTRIES OF PHI-BASE (NON-SQUARE SHAPE) COLOURED BY PHENOTYPE, RELATIONS: ORTHOLOGOUS (DARKER) AND PARALOGOUS (LIGHTER) 147

FIGURE 56 MIXED PHENOTYPE CLUSTER FOR *FUSARIUM GRAMINEARUM* WITH COLOUR LEGEND DENOTING THE HOST RESPONSE PHENOTYPE ON THE RIGHT. THE STAR SHAPE INDICATES RESPONSES FOR ANIMAL PATHOGENS AND CIRCLE SHAPE FOR PLANT PATHOGENS. 148

FIGURE 57 EXAMPLE OF A MORE HOMOGENEOUS CLUSTER FOR *FUSARIUM GRAMINEARUM* WITH COLOUR LEGEND AND NODE SHAPES AS ABOVE FIGURE 149

FIGURE 58 INTEGRATION PIPELINE FOR CONSENSUS METABOLIC NETWORK BETWEEN ARACYC AND KEGG, REUSING PREVIOUSLY PRESENTED INTEGRATION METHODS: ACCESSION BASED MAPPING, SYNONYM (NAME BASED) MAPPING AND STRUCTALIGN. ADDITIONALLY THE GRAPH-PATTERN MAPPING (HIGLIGHTED) AND SBML EXPORT (HIGHLIGHTED) HAVE BEEN INTRODUCED. 153

FIGURE 59 EXAMPLE FOR GRAPH-PATTERN MAPPING. CONCEPT PAIRS 1-4 AND 2-5 HAVE PREVIOUSLY BEEN IDENTIFIED AS BEING EQUIVALENT BETWEEN THE TWO DATABASES DB1 AND DB2. DB2 IS STRUCTURAL DIFFERENT FROM DB1; IT HAS ADDITIONAL ENZYME COMPLEX INFORMATION AS CONCEPT 6 (TRIANGLE). GRAPH-PATTERN MAPPING NOW TRIES TO INFER THE DASHED RELATION BETWEEN REACTION CONCEPTS 3 AND 7 (SQUARES). 155

FIGURE 60 EXCERPT OF KEGG (LOWER PART) TO ARACYC (UPPER PART) MAPPING SHOWING THE STRUCTURAL DIFFERENCES BETWEEN THE TWO DATABASES IN THE MIDDLE PART (ARACYC DOES REPRESENT PROTEIN COMPLEX CPLXQT-6, KEGG DOES NOT) 157

FIGURE 61 EXAMPLE: GLUTATHIONE METABOLISM IN BOTH ARACYC AND KEGG. ALL DARKER NODES ARE COMMON TO BOTH ARACYC AND KEGG. CIRCLES ARE METABOLITES; TRIANGLES REPRESENT ENZYMES IN THE INTEGRATED DATA. 158

FIGURE 62 EXCERPT SBML EXPORT OF GLUTATHIONE METABOLISM NETWORK SHOWING ALL REACTANTS ("LISTOFSPECIES") AND PARTS OF ONE REACTION LABELLED "R1". 159

FIGURE 63 SBML LOADED INTO CELLDESIGNER. EXPERT KNOWLEDGE IS REQUIRED TO ASSIGN INITIAL CONCENTRATIONS OF METABOLITES AND CONVERSION RATES TO THE NODES AND EDGES IN THE NETWORK. 160

FIGURE 64 FOAF PARSER (HIGHLIGHTED) WAS ADDED TO THE ONDEX SYSTEM. EXISTING DATA INTEGRATION AND ANALYSIS METHODS COULD BE REUSED. 162

FIGURE 65 META-GRAPH: SUBSET OF THE FRIEND-OF-A-FRIEND (FOAF) DATA REPRESENTED IN ONDEX. 164

Table of figures

FIGURE 66 OVERVIEW OF DIFFERENT RDF CLASSES AND ATTRIBUTES IN FOAF.. 164
FIGURE 67 RESULT OF THE IMPORT OF FIVE FOAF RDF FILES INTO ONDEX. CONCEPTS FOR PERSON (TRIANGLE) ARE CONNECTED TO EACH OTHER BY RELATIONS OF TYPE "KNOWS" (ARROWS)... 165
FIGURE 68 DARK ARROWS: BETWEEN CONCEPTS FOR PERSONS IDENTIFIED AS EQUIVALENT ACROSS THE PREVIOUSLY FIVE DIFFERENT CLUSTERS OF FOAF DATA USING ACCESSION BASED MAPPING. INSET: SINGLE PERSON "AVINASH SHENOI", OCCURRING TWICE. THESE OCCURRENCES WERE IDENTIFIED TO BE EQUIVALENT ("EQU")... 166
FIGURE 69 RESULT OF COLLAPSING EQUIVALENT CONCEPTS TOGETHER. ONLY ONE RELATION TYPE "KNOWS" IS PRESENT. 167
FIGURE 70 BETWEENNESS CENTRALITY APPLIED TO SOCIAL NETWORK. LEFT PANEL: BETWEENNESS CENTRALITY SCORES. SIZE OF A NODE IS PROPORTIONAL TO ITS SCORE. TOP FOUR SCORES: TIM FININ (1.0), ANUPAM JOSH (0.49), YELENA YESHA (0.08) AND YUN PENG (0.07).......... 168
FIGURE 71 EXAMPLE OF A SEMANTIC NETWORK DEPICTING CLASSIFICATION OF DIFFERENT MAMMALS... 202
FIGURE 72 FRAMEWORK FOR DISCUSSING DOMAIN MODELLING................... 204

REFERENCES

[1] BBSRC. (2007). Systems biology. [Corporate brochures]. Available: http://www.bbsrc.ac.uk/publications/corporate/systems-biology.aspx

[2] J. Köhler, J. Baumbach, J. Taubert, M. Specht, A. Skusa, A. Ruegg, C. Rawlings, P. Verrier, and S. Philippi, "Graph-based analysis and visualization of experimental results with ONDEX," *Bioinformatics*, vol. 22, pp. 1383-90, Jun 1 2006.

[3] M. Gaylord, J. Calley, H. Qiang, E. W. Su, and B. Liao, "A flexible integration and visualisation system for biomarker discovery," *Applied bioinformatics*, vol. 5, pp. 219-23, 2006.

[4] H. P. Fischer, "Towards quantitative biology: integration of biological information to elucidate disease pathways and to guide drug discovery," *Biotechnology annual review*, vol. 11, pp. 1-68, 2005.

[5] J. Köhler, C. Rawlings, P. Verrier, R. Mitchell, A. Skusa, A. Ruegg, and S. Philippi, "Linking experimental results, biological networks and sequence analysis methods using Ontologies and Generalised Data Structures," *In Silico Biol*, vol. 5, pp. 33-44, 2005.

[6] J. Taubert, M. Hindle, A. Lysenko, J. Weile, J. Köhler, and C. J. Rawlings, "Linking Life Sciences Data Using Graph-Based Mapping," presented at the Proceedings of the 6th International Workshop on Data Integration in the Life Sciences, Manchester, UK, 2009.

[7] J. Taubert, K. P. Sieren, M. Hindle, B. Hoekman, R. Winnenburg, S. Philippi, C. Rawlings, and J. Köhler, "The OXL format for the exchange of integrated datasets," *Journal of Integrative Bioinformatics*, vol. 4, 2007.

[8] M. Atkinson and D. De Roure, "Metaphors, Utopia and Reality," in *OpenWetWare* vol. 2010, D. DeRoure, Ed., ed, 2009.

[9] C. Goble and R. Stevens, "State of the nation in data integration for bioinformatics," *J Biomed Inform*, vol. 41, pp. 687-93, Oct 2008.

[10] D. B. Searls, "Data integration: challenges for drug discovery," *Nat Rev Drug Discov*, vol. 4, pp. 45-58, Jan 2005.

[11] L. Wong, "Technologies for integrating biological data," *Brief Bioinform*, vol. 3, pp. 389-404, Dec 2002.

[12] J. Köhler, "Integration of Life Science Databases," *Drugs Discovery Today: BioSilico*, vol. 2, pp. 61-69, 1 March 2004 2004.

[13] A. Ng, B. Bursteinas, Q. Gao, E. Mollison, and M. Zvelebil, "Resources for integrative systems biology: from data through databases to networks and dynamic system models," *Briefings in Bioinformatics*, vol. 7, pp. 318-330, 2006.

[14] F. Bry and P. Kröger, "A Computational Biology Database Digest: Data, Data Analysis, and Data Management," *Distrib. Parallel Databases*, vol. 13, pp. 7-42, 2003.

[15] T. Etzold, A. Ulyanov, and P. Argos, "SRS: information retrieval system for molecular biology data banks," *Methods Enzymol*, vol. 266, pp. 114-28, 1996.

References

[16] G. D. Schuler, J. A. Epstein, H. Ohkawa, and J. A. Kans, "Entrez: molecular biology database and retrieval system," *Methods Enzymol,* vol. 266, pp. 141-62, 1996.

[17] P. Kersey, L. Bower, L. Morris, A. Horne, R. Petryszak, C. Kanz, A. Kanapin, U. Das, K. Michoud, *et al.*, "Integr8 and Genome Reviews: integrated views of complete genomes and proteomes," *Nucleic Acids Res,* vol. 33, pp. D297-302, Jan 1 2005.

[18] C. Hedeler, H. M. Wong, M. J. Cornell, I. Alam, D. M. Soanes, M. Rattray, S. J. Hubbard, N. J. Talbot, S. G. Oliver, *et al.*, "e-Fungi: a data resource for comparative analysis of fungal genomes," *BMC Genomics,* vol. 8, p. 426, 2007.

[19] M. Cornell, N. W. Paton, C. Hedeler, P. Kirby, D. Delneri, A. Hayes, and S. G. Oliver, "GIMS: an integrated data storage and analysis environment for genomic and functional data," *Yeast,* vol. 20, pp. 1291-306, Nov 2003.

[20] S. P. Shah, Y. Huang, T. Xu, M. M. Yuen, J. Ling, and B. F. Ouellette, "Atlas - a data warehouse for integrative bioinformatics," *BMC Bioinformatics,* vol. 6, p. 34, 2005.

[21] S. Trissl, K. Rother, H. Muller, T. Steinke, I. Koch, R. Preissner, C. Frommel, and U. Leser, "Columba: an integrated database of proteins, structures, and annotations," *BMC Bioinformatics,* vol. 6, p. 81, 2005.

[22] *IBM Websphere Information Integrator.* Available: http://www.ibm.com/businesscenter/smb/uk/en/lifesciences

[23] *GMOD - Generic Model Organism Database.* Available: http://gmod.org

[24] S. Durinck, Y. Moreau, A. Kasprzyk, S. Davis, B. De Moor, A. Brazma, and W. Huber, "BioMart and Bioconductor: a powerful link between biological databases and microarray data analysis," *Bioinformatics,* vol. 21, pp. 3439-40, Aug 15 2005.

[25] T. J. Lee, Y. Pouliot, V. Wagner, P. Gupta, D. W. Stringer-Calvert, J. D. Tenenbaum, and P. D. Karp, "BioWarehouse: a bioinformatics database warehouse toolkit," *BMC Bioinformatics,* vol. 7, p. 170, 2006.

[26] A. Birkland and G. Yona, "BIOZON: a system for unification, management and analysis of heterogeneous biological data," *BMC Bioinformatics,* vol. 7, p. 70, 2006.

[27] R. Stevens, P. Baker, S. Bechhofer, G. Ng, A. Jacoby, N. W. Paton, C. A. Goble, and A. Brass, "TAMBIS: transparent access to multiple bioinformatics information sources," *Bioinformatics,* vol. 16, pp. 184-5, Feb 2000.

[28] S. Y. Chung and L. Wong, "Kleisli: a new tool for data integration in biology," *Trends Biotechnol,* vol. 17, pp. 351-5, Sep 1999.

[29] *Medical Integrator.* Available: http://www.medicel.com

[30] *ComparaGrid.* Available: http://www.comparagrid.org

[31] D. De Roure, C. Goble, and R. Stevens, "The design and realisation of the Virtual Research Environment for social sharing of workflows," *Future Generation Computer Systems,* vol. 25, pp. 561-567, 2009.

[32] *InforSense.* Available: http://www.inforsense.com

[33] *Pipeline Pilot.* Available: http://www.scitegic.com

References

[34] T. Oinn, M. Addis, J. Ferris, D. Marvin, M. Senger, M. Greenwood, T. Carver, K. Glover, M. R. Pocock, et al., "Taverna: a tool for the composition and enactment of bioinformatics workflows," *Bioinformatics*, vol. 20, pp. 3045-54, Nov 22 2004.

[35] R. D. Dowell, R. M. Jokerst, A. Day, S. R. Eddy, and L. Stein, "The distributed annotation system," *BMC Bioinformatics*, vol. 2, p. 7, 2001.

[36] P. Jones, N. Vinod, T. Down, A. Hackmann, A. Kahari, E. Kretschmann, A. Quinn, D. Wieser, H. Hermjakob, et al., "Dasty and UniProt DAS: a perfect pair for protein feature visualization," *Bioinformatics*, vol. 21, pp. 3198-9, Jul 15 2005.

[37] J. Kohler, S. Philippi, and M. Lange, "SEMEDA: ontology based semantic integration of biological databases," *Bioinformatics*, vol. 19, pp. 2420-7, Dec 12 2003.

[38] J. Kohler and S. Schulze-Kremer, "The semantic metadatabase (SEMEDA): ontology based integration of federated molecular biological data sources," *In Silico Biol*, vol. 2, pp. 219-31, 2002.

[39] A. Freier, R. Hofestadt, M. Lange, U. Scholz, and A. Stephanik, "BioDataServer: a SQL-based service for the online integration of life science data," *In Silico Biol*, vol. 2, pp. 37-57, 2002.

[40] A. Freier, R. Hofestädt, and T. Toepel, "Investigating the Effective Range of Agents by Using Integrative Modelling," in *Handbook of Toxicogenomics*, B. Univ.-Prof. Dr. Jürgen, Ed., ed, 2005, pp. 233-251.

[41] M. Baitaluk, X. Qian, S. Godbole, A. Raval, A. Ray, and A. Gupta, "PathSys: integrating molecular interaction graphs for systems biology," *BMC bioinformatics*, vol. 7, p. 55, 2006.

[42] J. Küntzer, T. Blum, A. Gerasch, C. Backes, A. Hildebrandt, M. Kaufmann, O. Kohlbacher, and H.-P. Lenhof, "BN++ - A Biological Information System," *Journal of Integrative Bioinformatics*, vol. 3, 2006.

[43] B. Smith, W. Ceusters, B. Klagges, J. Kohler, A. Kumar, J. Lomax, C. Mungall, F. Neuhaus, A. L. Rector, et al., "Relations in biomedical ontologies," *Genome Biol*, vol. 6, p. R46, 2005.

[44] M. Hsing, J. L. Bellenson, C. Shankey, and A. Cherkasov, "Modeling of cell signaling pathways in macrophages by semantic networks," *BMC Bioinformatics*, vol. 5, p. 156, Oct 19 2004.

[45] R. L. Griffith, "Three principles of representation for semantic networks," *ACM Transactions on Database Systems*, vol. 7, pp. 417-442, 1982.

[46] D. Lee, S. Kim, and Y. Kim, "BioCAD: an information fusion platform for bio-network inference and analysis," *BMC Bioinformatics*, vol. 8 Suppl 9, p. S2, 2007.

[47] J. Kuntzer, C. Backes, T. Blum, A. Gerasch, M. Kaufmann, O. Kohlbacher, and H. P. Lenhof, "BNDB - the Biochemical Network Database," *BMC Bioinformatics*, vol. 8, p. 367, 2007.

[48] L. J. Jensen, M. Kuhn, M. Stark, S. Chaffron, C. Creevey, J. Muller, T. Doerks, P. Julien, A. Roth, et al., "STRING 8--a global view on proteins and their functional interactions in 630 organisms," *Nucleic Acids Res*, vol. 37, pp. D412-6, Jan 2009.

References

[49] U. de Lichtenberg, L. J. Jensen, S. Brunak, and P. Bork, "Dynamic complex formation during the yeast cell cycle," *Science,* vol. 307, pp. 724-7, Feb 4 2005.

[50] S. Brohee, K. Faust, G. Lima-Mendez, O. Sand, R. Janky, G. Vanderstocken, Y. Deville, and J. van Helden, "NeAT: a toolbox for the analysis of biological networks, clusters, classes and pathways," *Nucleic Acids Res,* vol. 36, pp. W444-51, Jul 1 2008.

[51] T. Dwyer, H. Rolletschek, and F. Schreiber, "Representing experimental biological data in metabolic networks," presented at the Proceedings of the second conference on Asia-Pacific bioinformatics - Volume 29, Dunedin, New Zealand, 2004.

[52] E. Pennisi, "How Will Big Pictures Emerge From a Sea of Biological Data?," *Science,* vol. 309, p. 94, 2005.

[53] S. Philippi and J. Köhler, "Addressing the problems with life-science databases for traditional uses and systems biology," *Nature Reviews Genetics,* vol. 7, pp. 482-488, 2006.

[54] S. Ananiadou, D. B. Kell, and J. I. Tsujii, "Text mining and its potential applications in systems biology," *Trends Biotechnol,* vol. 24, pp. 571-579, Dec 2006.

[55] J. Augen, "Information technology to the rescue!," *Nat Biotechnol,* vol. 19 Suppl, pp. BE39-40, Jul 2001.

[56] R. Carel, "Practical data integration in biopharmaceutical research and development," *PharmaGenomics,* vol. 3, pp. 22–35, 2003.

[57] S. P. Gardner, "Ontologies and semantic data integration," *Drug Discov Today,* vol. 10, pp. 1001-7, Jul 15 2005.

[58] L. Stein, "Creating a bioinformatics nation," *Nature,* vol. 417, pp. 119-20, May 9 2002.

[59] L. Cardelli, "Type Systems," *ACM Comput. Surv.,* vol. 28, pp. 263-264, 1996.

[60] R. Apweiler, A. Bairoch, C. H. Wu, W. C. Barker, B. Boeckmann, S. Ferro, E. Gasteiger, H. Huang, R. Lopez, *et al.*, "UniProt: the Universal Protein knowledgebase," *Nucleic Acids Res,* vol. 32, pp. D115-9, Jan 1 2004.

[61] K. Eilbeck, S. E. Lewis, C. J. Mungall, M. Yandell, L. Stein, R. Durbin, and M. Ashburner, "The Sequence Ontology: a tool for the unification of genome annotations," *Genome Biol,* vol. 6, p. R44, 2005.

[62] B. Smith, M. Ashburner, C. Rosse, J. Bard, W. Bug, W. Ceusters, L. J. Goldberg, K. Eilbeck, A. Ireland, *et al.*, "The OBO Foundry: coordinated evolution of ontologies to support biomedical data integration," *Nat Biotechnol,* vol. 25, pp. 1251-5, Nov 2007.

[63] W3C. (2007). *OWL Features.* Available: http://www.w3.org/TR/owl-features/

[64] M. Brochhausen, "The derives_from relation in biomedical ontologies," *Stud Health Technol Inform,* vol. 124, pp. 769-74, 2006.

[65] H. Ogata, S. Goto, K. Sato, W. Fujibuchi, H. Bono, and M. Kanehisa, "KEGG: Kyoto Encyclopedia of Genes and Genomes," *Nucleic Acids Res,* vol. 27, pp. 29-34, Jan 1 1999.

[66] P. D. Karp, C. A. Ouzounis, C. Moore-Kochlacs, L. Goldovsky, P. Kaipa, D. Ahren, S. Tsoka, N. Darzentas, V. Kunin, *et al.*, "Expansion of the BioCyc

collection of pathway/genome databases to 160 genomes," *Nucleic Acids Res,* vol. 33, pp. 6083-9, 2005.

[67] J. L. Sussman, D. Lin, J. Jiang, N. O. Manning, J. Prilusky, O. Ritter, and E. E. Abola, "Protein Data Bank (PDB): database of three-dimensional structural information of biological macromolecules," *Acta Crystallogr D Biol Crystallogr,* vol. 54, pp. 1078-84, Nov 1 1998.

[68] J. M. Cherry, C. Ball, S. Weng, G. Juvik, R. Schmidt, C. Adler, B. Dunn, S. Dwight, L. Riles, et al., "Genetic and physical maps of Saccharomyces cerevisiae," *Nature,* vol. 387, pp. 67-73, May 29 1997.

[69] K. Tipton and S. Boyce, "History of the enzyme nomenclature system," *Bioinformatics,* vol. 16, pp. 34-40, Jan 2000.

[70] A. Bairoch, "The ENZYME database in 2000," *Nucleic Acids Res,* vol. 28, pp. 304-5, Jan 1 2000.

[71] M. Ashburner, C. A. Ball, J. A. Blake, D. Botstein, H. Butler, J. M. Cherry, A. P. Davis, K. Dolinski, S. S. Dwight, et al., "Gene ontology: tool for the unification of biology. The Gene Ontology Consortium," *Nat Genet,* vol. 25, pp. 25-9, May 2000.

[72] M. Kanehisa, S. Goto, S. Kawashima, Y. Okuno, and M. Hattori, "The KEGG resource for deciphering the genome," *Nucleic Acids Res,* vol. 32, pp. D277-80, Jan 1 2004.

[73] L. A. Mueller, P. Zhang, and S. Y. Rhee, "AraCyc: a biochemical pathway database for Arabidopsis," *Plant Physiol,* vol. 132, pp. 453-60, Jun 2003.

[74] B. Smith, "Beyond Concepts: Ontology as Reality Representation," in *Proceedings of FOIS,* 2004.

[75] M. J. Schuemie, B. Mons, M. Weeber, and J. A. Kors, "Evaluation of techniques for increasing recall in a dictionary approach to gene and protein name identification," *J Biomed Inform,* vol. 40, pp. 316-24, Jun 2007.

[76] D. Knuth, "Section 6.2.3: Balanced Trees," in *The Art of Computer Programming.* vol. Volume 3: Sorting and Searching, Third Edition ed: Addison-Wesley, 1997, pp. 458-481.

[77] W. R. Pearson, "Rapid and sensitive sequence comparison with FASTP and FASTA," *Methods Enzymol,* vol. 183, pp. 63-98, 1990.

[78] E. L. Sonnhammer, S. R. Eddy, and R. Durbin, "Pfam: a comprehensive database of protein domain families based on seed alignments," *Proteins,* vol. 28, pp. 405-20, Jul 1997.

[79] S. F. Altschul, T. L. Madden, A. A. Schaffer, J. Zhang, Z. Zhang, W. Miller, and D. J. Lipman, "Gapped BLAST and PSI-BLAST: a new generation of protein database search programs," *Nucleic Acids Res,* vol. 25, pp. 3389-402, Sep 1 1997.

[80] R. Durbin, S. Eddy, A. Krogh, and G. Mitchison, "The theory behind profile HMMs," in *Biological sequence analysis: probabilistic models of proteins and nucleic acids,* ed: Cambridge University Press, 1998.

[81] C. Goutte and E. Gaussier, "A Probabilistic Interpretation of Precision, Recall and F-score, with Implication for Evaluation," in *European Colloquium on IR Research (ECIR'05),* 2005, pp. 345-359.

References

[82] M. L. Green and P. D. Karp, "The outcomes of pathway database computations depend on pathway ontology," *Nucl. Acids Res.,* vol. 34, pp. 3687-3697, 2006.

[83] R. E. Buntrock, "Chemical registries--in the fourth decade of service," *J Chem Inf Comput Sci,* vol. 41, pp. 259-63, Mar-Apr 2001.

[84] D. Meinke, "Genetic nomenclature guide. Arabidopsis thaliana," *Trends in genetics,* pp. 22-3, Mar 1995.

[85] T. K. Baldwin, R. Winnenburg, M. Urban, C. Rawlings, J. Köhler, and K. E. Hammond-Kosack, "PHI-base provides insights into generic and novel themes of pathogenicity," *Molecular Plant-Microbe Interactions,* vol. 19, pp. 1451-62, 2006.

[86] R. Winnenburg, T. K. Baldwin, M. Urban, C. Rawlings, J. Köhler, and K. E. Hammond-Kosack, "PHI-base: A new database for Pathogen Host Interactions," *Nucleic Acids Res,* vol. 34, Database issue, pp. D459-64, Jan 1 2006.

[87] J. Köhler, K. Munn, A. Rüegg, A. Skusa, and B. Smith, "Quality Control for Terms and Definitions in Ontologies and Taxonomies," *BMC Bioinformatics,* vol. 7, p. 212, 2006.

[88] L. Zhang and J.-G. Gu, "Ontology based semantic mapping architecture," in *Fourth International Conference on Machine Learning and Cybernetics,* 2005.

[89] DoubleTwist Inc. (2001). *AGAVE: architecture for genomic annotation, visualization and exchange.* Available: http://www.agavexml.org

[90] G. Bader and M. Cary. (2005). *BioPAX - Biological Pathways Exchange Language.* Available: http://www.biopax.org/release/biopax-level2-documentation.pdf

[91] D. Fenyo, "The Biopolymer Markup Language," *Bioinformatics,* vol. 15, pp. 339-40, Apr 1999.

[92] LabBook Inc. (2002). *BSML (bioinformatics sequence markup language) 3.1.* Available: http://www.bsml.org

[93] C. M. Lloyd, M. D. Halstead, and P. F. Nielsen, "CellML: its future, present and past," *Prog Biophys Mol Biol,* vol. 85, pp. 433-50, Jun-Jul 2004.

[94] Y. M. Liao and H. Ghanadan, "The chemical markup language," *Anal Chem,* vol. 74, pp. 389A-390A, Jul 1 2002.

[95] P. T. Spellman, M. Miller, J. Stewart, C. Troup, U. Sarkans, S. Chervitz, D. Bernhart, G. Sherlock, C. Ball, *et al.,* "Design and implementation of microarray gene expression markup language (MAGE-ML)," *Genome Biol,* vol. 3, p. RESEARCH0046, Aug 23 2002.

[96] C. F. Taylor, N. W. Paton, K. L. Garwood, P. D. Kirby, D. A. Stead, Z. Yin, E. W. Deutsch, L. Selway, J. Walker, *et al.,* "A systematic approach to modeling, capturing, and disseminating proteomics experimental data," *Nat Biotechnol,* vol. 21, pp. 247-54, Mar 2003.

[97] D. Hanisch, R. Zimmer, and T. Lengauer, "ProML--the protein markup language for specification of protein sequences, structures and families," *In Silico Biol,* vol. 2, pp. 313-24, 2002.

[98] H. Hermjakob, L. Montecchi-Palazzi, G. Bader, J. Wojcik, L. Salwinski, A. Ceol, S. Moore, S. Orchard, U. Sarkans, *et al.,* "The HUPO PSI's molecular

interaction format--a community standard for the representation of protein interaction data," *Nat Biotechnol*, vol. 22, pp. 177-83, Feb 2004.
[99] M. Hucka, A. Finney, B. J. Bornstein, S. M. Keating, B. E. Shapiro, J. Matthews, B. L. Kovitz, M. J. Schilstra, A. Funahashi, *et al.*, "Evolving a lingua franca and associated software infrastructure for computational systems biology: the Systems Biology Markup Language (SBML) project," *Syst Biol (Stevenage)*, vol. 1, pp. 41-53, Jun 2004.
[100] H. W. Mewes, D. Frishman, K. F. Mayer, M. Munsterkotter, O. Noubibou, P. Pagel, T. Rattei, M. Oesterheld, A. Ruepp, *et al.*, "MIPS: analysis and annotation of proteins from whole genomes in 2005," *Nucleic Acids Res*, vol. 34, pp. D169-72, Jan 1 2006.
[101] M. Hucka, A. Finney, H. M. Sauro, H. Bolouri, J. C. Doyle, H. Kitano, A. P. Arkin, B. J. Bornstein, D. Bray, *et al.*, "The systems biology markup language (SBML): a medium for representation and exchange of biochemical network models," *Bioinformatics*, vol. 19, pp. 524-31, Mar 1 2003.
[102] B. Yu, M. Dai, S. J. Watson, and F. Meng, "A Database-Linked Diagram Editor for Biological Concepts," in *The 2005 International Conference on Mathematics and Engineering Techniques in Medicine and Biological Sciences*, 2005, pp. 281-286.
[103] S. P. Kumar and J. C. Feidler, "BioSPICE: a computational infrastructure for integrative biology," *Omics*, vol. 7, p. 225, Fall 2003.
[104] B. E. Shapiro, A. Levchenko, E. M. Meyerowitz, B. J. Wold, and E. D. Mjolsness, "Cellerator: extending a computer algebra system to include biochemical arrows for signal transduction simulations," *Bioinformatics*, vol. 19, pp. 677-8, Mar 22 2003.
[105] P. Shannon, A. Markiel, O. Ozier, N. S. Baliga, J. T. Wang, D. Ramage, N. Amin, B. Schwikowski, and T. Ideker, "Cytoscape: a software environment for integrated models of biomolecular interaction networks," *Genome Res*, vol. 13, pp. 2498-504, Nov 2003.
[106] I. Goryanin, T. C. Hodgman, and E. Selkov, "Mathematical simulation and analysis of cellular metabolism and regulation," *Bioinformatics*, vol. 15, pp. 749-58, Sep 1999.
[107] M. Tomita, K. Hashimoto, K. Takahashi, T. S. Shimizu, Y. Matsuzaki, F. Miyoshi, K. Saito, S. Tanida, K. Yugi, *et al.*, "E-CELL: software environment for whole-cell simulation," *Bioinformatics*, vol. 15, pp. 72-84, Jan 1999.
[108] P. Mendes, "GEPASI: a software package for modelling the dynamics, steady states and control of biochemical and other systems," *Comput Appl Biosci*, vol. 9, pp. 563-71, Oct 1993.
[109] P. D. Karp, S. Paley, and P. Romero, "The Pathway Tools software," *Bioinformatics*, vol. 18 Suppl 1, pp. S225-32, 2002.
[110] E. Demir, O. Babur, U. Dogrusoz, A. Gursoy, G. Nisanci, R. Cetin-Atalay, and M. Ozturk, "PATIKA: an integrated visual environment for collaborative construction and analysis of cellular pathways," *Bioinformatics*, vol. 18, pp. 996-1003, Jul 2002.
[111] N. Le Novere and T. S. Shimizu, "STOCHSIM: modelling of stochastic biomolecular processes," *Bioinformatics*, vol. 17, pp. 575-6, Jun 2001.

References

[112] J. C. Schaff, B. M. Slepchenko, Y. S. Choi, J. Wagner, D. Resasco, and L. M. Loew, "Analysis of nonlinear dynamics on arbitrary geometries with the Virtual Cell," *Chaos,* vol. 11, pp. 115-131, Mar 2001.

[113] Z. Hu, J. Mellor, J. Wu, T. Yamada, D. Holloway, and C. Delisi, "VisANT: data-integrating visual framework for biological networks and modules," *Nucleic Acids Res,* vol. 33, pp. W352-7, Jul 1 2005.

[114] P. Ion and R. Miner. (1999). *Mathematical Markup Language (MathML) 1.01 Specification.* Available: http://www.w3.org/TR/REC-MathML/

[115] T. Bray, J. Paoli, and C. M. Sperberg-McQueen, "Extensible markup language," *World Wide Web J.,* vol. 2, pp. 29-66, 1997.

[116] F. Achard, G. Vaysseix, and E. Barillot, "XML, bioinformatics and data integration," *Bioinformatics,* vol. 17, pp. 115-25, Feb 2001.

[117] A. Ruttenberg, J. A. Rees, and J. S. Luciano, "Experience using OWL DL for the exchange of biological pathway information," in *OWL Experiences and Directions,* 2005.

[118] E. Wingender, P. Dietze, H. Karas, and R. Knuppel, "TRANSFAC: a database on transcription factors and their DNA binding sites," *Nucleic Acids Res,* vol. 24, pp. 238-41, Jan 1 1996.

[119] F. Schacherer, C. Choi, U. Gotze, M. Krull, S. Pistor, and E. Wingender, "The TRANSPATH signal transduction database: a knowledge base on signal transduction networks," *Bioinformatics,* vol. 17, pp. 1053-7, Nov 2001.

[120] Oracle-Corp., "Berkeley DB Java Edition, Direct Persistence Layer Basics," ed. Redwood Shores, 2006, p. 15.

[121] H. Yu and E. Agichtein, "Extracting synonymous gene and protein terms from biological literature," *Bioinformatics,* vol. 19 Suppl 1, pp. i340-9, 2003.

[122] S. Schulz, E. Beisswanger, J. Wermter, and U. Hahn, "Towards an upper level ontology for molecular biology," *AMIA Annu Symp Proc,* pp. 694-8, 2006.

[123] H. Stenzhorn, E. Beisswanger, and S. Schulz, "Towards a top-domain ontology for linking biomedical ontologies," *Stud Health Technol Inform,* vol. 129, pp. 1225-9, 2007.

[124] D. K. Button, K. M. Gartland, L. D. Ball, L. Natanson, J. S. Gartland, and G. D. Lyon, "DRASTIC--INSIGHTS: querying information in a plant gene expression database," *Nucleic Acids Res,* vol. 34, pp. D712-6, Jan 1 2006.

[125] C. E. Lipscomb, "Medical Subject Headings (MeSH)," *Bull Med Libr Assoc,* vol. 88, pp. 265-6, Jul 2000.

[126] T. Greenhalgh, "How to read a paper. The Medline database," *Bmj,* vol. 315, pp. 180-3, Jul 19 1997.

[127] L. Stromback, V. Jakoniene, H. Tan, and P. Lambrix, "Representing, storing and accessing molecular interaction data: a review of models and tools," *Brief Bioinform,* vol. 7, pp. 331-8, Dec 2006.

[128] C. Blaschke, M. A. Andrade, C. Ouzounis, and A. Valencia, "Automatic extraction of biological information from scientific text: protein-protein interactions," *Proc Int Conf Intell Syst Mol Biol,* pp. 60-7, 1999.

[129] Y. Liu, S. B. Navathe, A. Pivoshenko, V. G. Dasigi, R. Dingledine, and B. J. Ciliax, "Text analysis of MEDLINE for discovering functional relationships

among genes: evaluation of keyword extraction weighting schemes," *Int J Data Min Bioinform,* vol. 1, pp. 88-110, 2006.

[130] A. Eliens, *Principles of Object-Oriented Software Development* Addison-Wesley, 1995.

[131] G. Bracha. (2004). Generics in the Java Programming Language. Available: http://java.sun.com/j2se/1.5/pdf/generics-tutorial.pdf

[132] *Class BitSet.* Available: http://java.sun.com/javase/6/docs/api/java/util/BitSet.html

[133] *Interface Iterable.* Available: http://java.sun.com/javase/6/docs/api/java/lang/Iterable.html

[134] G. Boole, "The Calculus of Logic," *Cambridge and Dublin Mathematical Journal,* vol. III, pp. 183-98, 1848.

[135] R. Pesch, A. Lysenko, M. Hindle, K. Hassani-Pak, R. Thiele, C. Rawlings, J. Köhler, and J. Taubert, "Graph-based sequence annotation using a data integration approach," *Journal of Integrative Bioinformatics,* vol. 5, 2008.

[136] S. Salzberg, "Genome re-annotation: a wiki solution?," *Genome Biol,* vol. 8, p. 102, 2007.

[137] S. Kiritchenko, S. Matwin, and A. F. Famili, "Functional annotation of genes using hierarchical text categorization," *Proc. BioLINK SIG Meeting on Text Data Mining at ISMB'05,* 2005.

[138] N. Bluthgen, K. Brand, B. Cajavec, M. Swat, H. Herzel, and D. Beule, "Biological profiling of gene groups utilizing Gene Ontology," *Genome Inform,* vol. 16, pp. 106-15, 2005.

[139] M. Remm, C. E. Storm, and E. L. Sonnhammer, "Automatic clustering of orthologs and in-paralogs from pairwise species comparisons," *J Mol Biol,* vol. 314, pp. 1041-52, Dec 14 2001.

[140] H. Y. Yu, J. A. Seo, J. E. Kim, K. H. Han, W. B. Shim, S. H. Yun, and Y. W. Lee, "Functional analyses of heterotrimeric G protein G alpha and G beta subunits in Gibberella zeae," *Microbiology,* vol. 154, pp. 392-401, Feb 2008.

[141] M. Y. Galperin and G. R. Cochrane, "Nucleic Acids Research annual Database Issue and the NAR online Molecular Biology Database Collection in 2009," *Nucleic Acids Res,* vol. 37, pp. D1-4, Jan 2009.

[142] A. C. Weaver and B. B. Morrison, "Social Networking," *Computer,* vol. 41, pp. 97-100, 2008.

[143] D. DiNucci, "Fragmented Future," *Print: Design & New Media,* vol. 53, p. 32, 1999.

[144] U. Brandes, "A Faster Algorithm for Betweenness Centrality," *Journal of Mathematical Sociology* vol. 25, pp. 163-177, 2001.

[145] U. Brandes, "A Faster Algorithm for Betweenness Centrality," *Journal of Mathematical Socio,* vol. 25, pp. 163-177, 2001.

[146] M. J. Herrgard, N. Swainston, P. Dobson, W. B. Dunn, K. Y. Arga, M. Arvas, N. Bluthgen, S. Borger, R. Costenoble, *et al.*, "A consensus yeast metabolic network reconstruction obtained from a community approach to systems biology," *Nat Biotechnol,* vol. 26, pp. 1155-60, Oct 2008.

References

[147] Y. C. Kim, Y. Yamaguchi, N. Kondo, H. Masutani, and J. Yodoi, "Thioredoxin-dependent redox regulation of the antioxidant responsive element (ARE) in electrophile response," *Oncogene,* vol. 22, pp. 1860-5, Mar 27 2003.

[148] S. Muggelton, "Inductive logic programming," *New Generation Computing,* vol. 8, 1990.

[149] J.-P. Rodrigue, C. Comtois, and B. Slack, *The Geography of Transport Systems*: New York: Routledge, 2006.

[150] R. Johnsonbaugh, *Discrete Mathematics*: Prentice Hall, 2008.

[151] G. Booch, *Object-Oriented Analysis and Design with Applications (3rd Edition)*: Addison Wesley Longman Publishing Co., Inc., 2004.

[152] OMG. (2002). Unified Modeling Language 2.0. Available: http://www.uml.org/

[153] J. Rumbaugh and M. Blaha, *Object-oriented modeling and design*: Prentice Hall IPE, 1991.

[154] T. Elrad, R. E. Filman, and A. Bader, "Aspect-oriented programming: Introduction," *Commun. ACM,* vol. 44, pp. 29-32, 2001.

[155] A. Newell, *The knowledge level*: American Association for Artificial Intelligence, 1988.

[156] J. Sowa, *Knowledge Representation: Logical, Philosophical, and Computational Foundations*: Course Technology, 1999.

[157] R. Davis, H. Shrobe, and P. Szolovits, "What Is a Knowledge Representation?," *AI Magazine,* vol. 14, pp. 17-33, 1993.

[158] A. Lysenko, M. M. Hindle, J. Taubert, M. Saqi, and C. J. Rawlings, "Data integration for plant genomics--exemplars from the integration of Arabidopsis thaliana databases," *Brief Bioinform,* vol. 10, pp. 676-93, Nov 2009.

[159] T. Harley, *The Psychology of Language - From Data to Theory*: Psychology Press, 1995.

[160] C. A. Welty, "An Integrated Representation for Software Development and Discovery," Rensselaer Polytechnic Institute, 1995.

[161] B. Hjorland and H. Albrechtsen, "Toward a new horizon in information science: Domain-analysis," *Journal of the American Society for Information Science,* vol. 46, pp. 400-425, 1995.

[162] I. Neil, B. W. Gerald, and A. Guillermo, "Domain modeling for software engineering," presented at the Proceedings of the 13th international conference on Software engineering, Austin, Texas, United States, 1991.

[163] E. Schoen, "Active Assistance for Domain Modeling," *Knowledge-Based Software Engineering Conference Proceedings,* pp. 26-35, 1991.

Die VDM Verlagsservicegesellschaft sucht für wissenschaftliche Verlage abgeschlossene und herausragende

Dissertationen, Habilitationen, Diplomarbeiten, Master Theses, Magisterarbeiten usw.

für die kostenlose Publikation als Fachbuch.

Sie verfügen über eine Arbeit, die hohen inhaltlichen und formalen Ansprüchen genügt, und haben Interesse an einer honorarvergüteten Publikation?

Dann senden Sie bitte erste Informationen über sich und Ihre Arbeit per Email an *info@vdm-vsg.de*.

Sie erhalten kurzfristig unser Feedback!

VDM Verlagsservicegesellschaft mbH
Dudweiler Landstr. 99   Telefon +49 681 3720 174
D - 66123 Saarbrücken   Fax   +49 681 3720 1749
www.vdm-vsg.de

Die VDM Verlagsservicegesellschaft mbH vertritt

Printed by Books on Demand GmbH, Norderstedt / Germany